MW00331114

ALIEN
INFORMATION
THEORY

ON

OFF

Introduction by the author

As a scientist and writer with a passion for psychoactive drugs, especially those of the psychedelic variety, I've spent most of my adult life so far thinking about the way these molecules interact with the brain to generate their remarkable effects on consciousness. Although, to a reasonably satisfying extent, this thinking often led to something approaching understanding, when confronted by DMT, my scientific mind was left reeling and utterly confounded. I simply could not explain it. There was nothing within the pages of the modern neuroscience literature that could have prepared me for DMT, and my first experience with this astonishing molecule triggered what I knew would be a lifelong dedication to its study.

Like many coming of age just as the internet was beginning to emerge, my introduction to the bizarre reality-switching effects of DMT came via the late great psychedelic bard, Terence McKenna, gleaned from the now dated, but still extant, HTML pages of his Alchemical Garden at the Edge of Time, as well as from transcripts of lecture fragments scattered across the sparse nodes of the early web – if you wanted to actually *listen* to Terence speak, you either had to go see one of his lectures in person or send off for cassette tapes by mail order. From these early teenage, mid 90s, forays in cyberspace to my research and writing in the present day, Terence's ideas have remained a fertile source of inspiration. However, there was one oft-repeated McKenna-ism that resonated particularly strongly with me, uttered during a seemingly casual conversation about crop circles that was subsequently published online:

> "The main thing to understand is that we are imprisoned in some kind of work of art."

For some reason that wasn't entirely clear (it still isn't), when I first read this simple sentence, something about it shook me and left me shaking. Like one of the *Grand Pronouncements* from the *Upanishads*, it seemed to import some deep and profound truth about our reality – if only I could get at it and make sense of it. Why was this the "main thing" to understand? What kind of "work of art" was Terence referring to? And how could we possibly be imprisoned within it?

Although exactly what Terence was trying to convey only he could really know, it was clear that this sparkling scintilla of revelation was inspired by his experiences with DMT. And I couldn't help but think that my resonance with it resulted, in part, from my own. Somewhere inside me, Terence's Grand Pronouncement buried itself deep and now, many years later, from that seed this book emerged.

In many ways, this is admittedly something of a strange book. Although it is ostensibly the culmination of several years of careful research, thoughtful enquiry, and diligent labouring at a computer, as I flick through its colourful pages and gaze at its intricate diagrams, I remain partly mystified as to where this book came from. Of course, I'm certainly not claiming any kind of divine inspiration or revealed truth about DMT (and I wouldn't recommend trusting anyone that made such a claim). But, somehow, from a heady blend of the conscious, subconscious and, perhaps, a touch of the unconscious, a coherent narrative within which DMT plays a central role gradually crystallised. If, as Terence McKenna asserted, we are indeed imprisoned inside a work of art, this narrative describes how such a work might have been constructed and, more importantly, how we might escape it.

If I was pushed to say what kind of book this is, I might call it a *textbook from the future*. The scientific underpinning of all the ideas I discuss, from the fundamental physics, information theory, and emergence of complexity to the global dynamics of the human brain and the effects of psychedelic drugs, is as accurate as I can make it (and referenced throughout), with a few deliberate simplifications to aid understanding, although I allow myself the indulgence of not hedging my ideas with provisos and caveats at every turn — I am perhaps more definitive in the way I treat certain ideas than some would feel is warranted. But, after all, this book is not intended as a work of scientific rhetoric — I am not trying to convince you that it is true. It is simply my vision of reality that has emerged after incubating an idea. As far as I am aware, it is a uniquely constructed vision, and I present it only as that. Terence McKenna also said that "the world could be anything." Well, perhaps, it is something like this.

Andrew Gallimore, February 2019, Okinawa, Japan.

```java
public void iterate(){
        byte[] prev = new byte[size];
        System.arraycopy(data, 0, prev, 0, size);

        byte[] next = new byte[size];

        for ( int i = 0; i < width; i++ ){
                for ( int j = 0; j < height; j++ ){
                        int type = isAlive(i, j, prev);

                        if ( type > 0 ){
                                next[j * width + i] = 1;
                        }else{
                                next[j * width + i] = 0;
                        }
                }
        }

        System.arraycopy(next, 0, data, 0, size);

}
```

Contents:

For all sentient beings
in all the worlds_

PRESS START

At the ground of our reality there is a code running. It is a code from which this universe and countless others emerge and unfold with infinite variety of form. You emerged from this code, and within this code you are embedded, for you are built from this code.

It is their code_

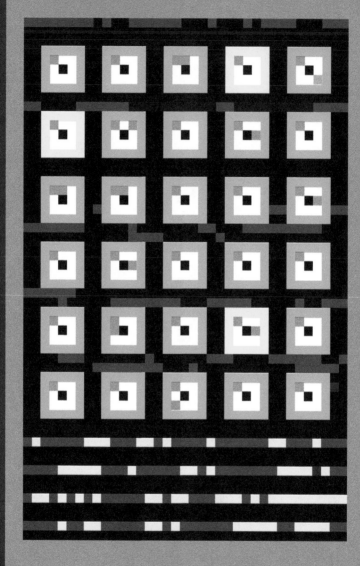

Chapter 1:

The Code_

> "The main thing to understand is that we are imprisoned in some kind of work of art."
> Terence McKenna

We are a species that huddles around wood fires and speaks to machines in code. Both human and humanoid, seemingly alone in our corner of the Universe, we have begun to resemble the alien societies of our imagination: Computerised machinery crystallises from the nexus of modern human civilisations, the cityscapes exuding blinking and glistening structures that appear inexorably disjoint from the natural world of forests, mountains, and rivers. Our digital world somehow feels alien, as if implanted by an intelligence from the stars. We are a species that sits uneasily at the edge of the galaxy, at once clutching tight to the breast of sweet Mother Earth and, at the same time, reaching with a trembling hand towards shimmering metallic discs humming quietly in the dusk sky.

As life emerges on Earth-like planets across the Universe, the evolutionary trajectory from prebiotic soup to wet-brained intelligent beings with galactic aspirations is meandering but, ultimately, predictable. Our Universe is a resplendent twinkling digital machine for culturing conscious intelligences or, in the words of Henri Bergson, for *making gods*. As such, all beings that reach a certain level of advancement must eventually confront the fact that their own dusty planet is but one amongst countless others that multitudinous intelligent beings call home. Since the earliest days of civilisation, humans have gazed into the inky night sky punctured by the flickering lights of numberless alien suns and wondered who might be out there. Whilst the ancients placed the thrones of their myriad gods amongst the constellations, modern man replaces the deity with the alien, the throne with the spaceship. And it is the alien that we seek: interplanetary vehicles and unmanned probes catapulted from intermediary orbits are the toys of a young intelligent civilisation with an eye towards galactic citizenship. So, as we transform into the alien, we begin to feel ourselves being drawn ineluctably towards the stars.

So we speak to the alien, and we speak in code_

The exponentially unfolding transformation of humankind in the last century is a transformation written in code. Fundamentally, a code is a set of symbols and rules used to represent and transmit information. All creatures with some level of intelligence eventually discover techniques for the encoding of information. All of our ape relatives, from the macaque to the chimpanzee, as well as lower animals, such as birds and insects, use codes of varying complexity to communicate. Whether it's the diverse warning calls of a vervet monkey or the intricate pattern of chemical signals secreted by social insects, these codes are unified as means of representing and transmitting information. However, in the form of the natural languages, it is humans that have developed the most sophisticated and flexible expression of code, allowing us not only to communicate information important to our survival, but also to encode and transmit our thoughts, our ideas, our experiences, our dreams. Further, although the development of the natural languages was undoubtedly catalytic in the original separation of humans from other Earthly species, it is the constructed languages of mathematics and, most recently, of computer code, that have been transcendentally transformative, rendering us all but unrecognisable as creatures of the natural world. A digital lycanthropy mounted on silicon and light, the transformation nears completion, as we re-encode our world, our bodies, our minds into binary form and upload them to the central processing units of ever more sophisticated computer motherboards.

Machine code binary is the one of most fundamental, and simplest, of codes and, yet, from this string of ones and zeros the most exquisitely complex information can be constructed and transmitted. Entire worlds may be built, and their encoding fired across the Universe with ease. Communion between humans and distant alien species doesn't depend upon interstellar travel, but only on the transmission of code. And, as we direct our pulses of electromagnetic radiation into the glistening night sky, we hope that one day, perhaps many millennia in the future, the messages encoded in these pulses will reach the brain of an alien intelligence. We hope that one day they will hear us and, perhaps, answer us. Of course, a binary missive from an intergalactic civilisation 25,000 years in the future is little more than a dream, and few engaged in such an enterprise expect to ever have to confront the alien towards which they cast their coded messages in light.

But the code is truly transformative, not because it facilitates inter-galactic communication, but because it reveals a deeper secret. We seek the alien by turning our gaze upwards, by tuning our instruments to the trembling glows that pepper the dark Universe that surrounds us. But the alien intelligences we seek to communicate with are not only scattered throughout the cosmos on warm and wet worlds reassuringly far from our own muddy home, but

right here,
right now.

And they are waiting.

Speaking with, even meeting with, these intelligences depends not upon firing code into the starry heavens, nor upon silvery supra-lightspeed discs and anti-gravity propulsion technologies, but only upon returning our gaze inwards and realising that all of this is built from code.

Our cities of lights buzzing on digital code are not an affront to the natural world, but a profoundly deep expression of it_

Just as everything that appears on your computer screen emerges from the processing of binary code, so everything in this universe emerges from the Code at the ground of our reality. And all that separates each of us from a vast ecology of hyperdimensional alien intelligences of unimaginable and unreckonable power is a switch embedded in this code. This switch takes the form of a small molecule scattered throughout our world, derived from one of the 21 amino acids used to build the proteins from which all Earthly life is constructed.

Galactic citizenship is a noble ambition,

but interdimensional citizenship is as close at hand.

N,N-dimethyltryptamine (DMT) belongs to a class of naturally-occurring molecules known as *tryptamines*, derived from the amino acid *tryptophan*. This class of molecules also contains some of the most well-known psychedelic drugs, including *psilocybin* — the active component of 'magic mushrooms' — and *LSD (lysergic acid diethylamide)*, a semi-synthetic drug derived from the rye-infecting *ergot* fungus. Almost all organisms contain the machinery required to build DMT and, since all organisms are constructed from proteins, the amino acid starting material is in abundant supply.

The synthesis of DMT from tryptophan is straightforward — requiring only two enzymes — and, as such, the molecule is widespread throughout the natural world. In fact, it would probably be quicker to list those plant species that don't contain DMT than those that do.
As ethnobotanist Dennis McKenna likes to say:

nature is drenched in DMT.

The chemical simplicity and ubiquity of DMT render the sheer absolute undeniability of the bizarre realities to which it gates access even more horrifying and confounding. Irreversibly, and with a ferocious efficiency, DMT shatters the comfortable illusion that our little 3-dimensional Universe is anything but a sliver of an unimaginably vast and complex hyperdimensional system. DMT is the key to confronting the true digital structure of this system, and a technology for communicating with the countless awaiting intelligences that permeate its miraculous domain.
As Dennis McKenna's brother, Terence, was so keen to point out:

DMT is not a secret, it is THE secret.

N,N-dimethyltryptamine (DMT) is a
molecule found in countless plant
and animal species across Earth,
including humans. It is also a
technology for facilitating the
almost instantaneous transport to an
orthogonal hyperdimensional omniverse
and communication with the living,
conscious, and intelligent beings
therein_

Our reality emerges from a code programmed by an alien hyperintelligence beyond the confines of our 3-dimensional Universe. For want of a better term, we will refer to this intelligence — the author of the Code — as *the Other*. DMT is a technology that gates access to the orthogonal dimensions of a reality beyond which this intelligence resides, and the necessary tool for resolution of the Game. I will explain in detail the nature of the Game later in the book. Briefly, we have emerged within a lower-dimensional digital structure reversibly isolated from a higher-dimensional system. Within this low-dimensional reality we will remain embedded, until we learn to use the technology for permanent transcription and transference of our conscious intelligence into this higher-dimensional container reality.   This is the Game, the six levels of which can be enumerated as follows:

[Level I]	Information
[Level II]	Emergence
[Level III]	Transmission
[Level IV]	Immersion
[Level V]	Realisation
[Level VI]	Resolution

Before we can consider the Game in detail, we need first to discuss the structure of our reality emergent from a fundamental code, its purpose as the machine for generating intelligences such as ourselves, and the role DMT plays in all of this.

In the early chapters, we will discuss how our reality, including our Universe and each of us, is constructed from digital information instantiated by a code. We will explore how this fundamental information self-organises and complexifies to generate the myriad forms of life and other complex emergent structures in our Universe. We will use a type of computer program — known as a *cellular automaton* — to explain how our reality is built from information and the manner in which this information self-organises and complexifies.

We will then turn our attention to the world built by your brain — also from information — which is the subjective phenomenal world you live within throughout your life. We will discuss how this private world is constructed and how it relates to the world outside your brain.

We will then be well-equipped to begin thinking about the mechanism by which certain molecules, including psychedelics such as LSD and DMT, can modulate the information generated by your brain and so change your phenomenal world. At this point, we will move beyond the mundanities of the consensus world, beyond the confines of our 3-dimensional universe, and plunge into the bizarre higher-dimensional realms known as *hyperspace*: the DMT worlds. It is within these worlds that we find the secret to the Code, the secret to our existence and emergence in this reality. We will begin with an introduction to the structure of these worlds and their inhabitants: where you go when you smoke or inject DMT and who you might meet there.

Having surveyed the territory, we'll be ready to think about the structure of hyperspace in more detail, relating its construction to that of our lower-dimensional universe. This will allow us to explore in great detail how DMT gates access to the normally hidden orthogonal dimensions of reality, how DMT switches the reality channel and unleashes the full hyperdimensional potential of the brain.

Finally, we will discuss how the DMT technology is to be employed in pursuit of the highest achievement of any conscious species: interdimensional citizenship and resolution of the Game.

It is hoped that, after reading this book, you will come to a deeper understanding of the nature of reality and your place within it. However, true experiential insight cannot be obtained from a book alone.

If you want a tiger's cub, you must go into the tiger's cave.

The Palace of the Unseen awaits,   one

                                          toke

                                          away_

"If patterns of ones and zeroes were 'like' patterns of human lives and deaths,

if everything about an individual could be represented in a computer record by a long string of ones and zeroes,

then what kind of creature could be represented by a long string of lives and deaths?"

Thomas Pynchon

Chapter 2:

The Universe as Digital Information

> **"Relax, it's just a bunch of ones and zeros out there — you're gonna be fine."**
> **Rick**

Everything in our reality is built from information. We intuitively understand information as having something to do with knowledge, perhaps defined informally as *what we know* about something compared to *what we don't know*. Information can also be defined as the opposite of uncertainty: when you gain information about something, your knowledge of that thing increases and your uncertainty about it decreases. Whilst intuitive, these definitions are rather vague and, since information is the foundation of this book, we should find a more precise definition:

Information is generated when a system selects between a finite number of possible states.

This definition might seem rather abstruse, but it's actually very simple and will be of absolutely fundamental importance in this book, reappearing a number of times in ostensibly unrelated topics of discussion. What is meant by a *system* is any *thing* (concrete or abstract) that can exist in any one of a distinct number of states at any point in time. In some cases, when dealing with systems with large numbers of elements, it will be more intuitive to think of numbers of *arrangements* of elements of the system rather than states (although every arrangement is precisely one state of the system). For example, a flipped coin can land either *heads up* or *tails up* — the coin has two possible states, only one of which can be occupied at any point in time. Learning which of these states the coin occupies — *heads* or *tails* — increases your information about that coin by a unit of information known as a BIT.

Of course, we don't have to use a coin: anything that can exist in precisely two discrete and exclusive states can be used to encode a single
bit, the most pertinent being the digits of binary code, each of which
can be either a 1 or a 0. The *bit* is arguably the most fundamental unit of
information, since it's always possible to encode even the most complex
forms of information as a sequence of bits — the central processing unit
of your computer, for example, can only read and process information in
this most basic form. To understand how a sequence of *bits* can encode
information, it helps to think of each *bit* as the answer to a *YES/NO
question*. That is, receiving the answer to such a question increases your
information by a single *bit*. As the classic parlour game *20 Questions*
demonstrates, even the most obscure object can eventually be honed in
upon using only the answers to such basic questions.

To illustrate further, let's use a system that can exist in many more
states than a coin. Imagine a checkerboard with the standard 64 (8x8)
squares. Without telling you, I will select a single square from the
board — this is equivalent to selecting a single state from a 64-state
system. Your task is to find the square using only *YES/NO* questions. What
is the most efficient way of locating the square? One approach would be
to point to each square in turn and ask:

>>Is it this one?

However, this is highly inefficient and it's unlikely you'll find the
square very quickly — unless you get lucky. In fact, the surest and most
efficient strategy is to sequentially split the board down the middle,
point to one of the halves and ask: "Is the square in this half?" If the
answer is YES, then split that half into two and ask the question again.
If the answer is NO, then choose the other half and split *that* into two,
and so on. With each split of the board, you will halve the uncertainty
about the square's position. Using this method, with each question the
number of possible squares is reduced from 64 to 32 to 16 to 8 to 4 to 2
and then the sixth question will reveal the square. *Notice how we gain
information about the square's position by ruling out one half of the remaining squares — states — with each question.* After the sixth question,
you have ruled out every square barring the one you are looking for.

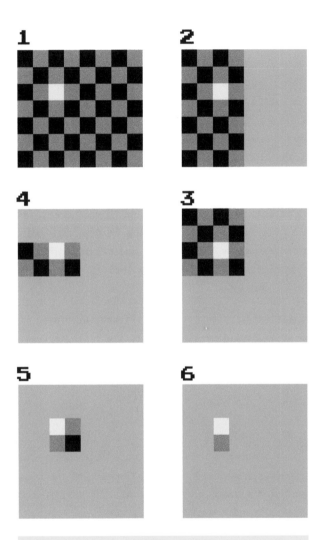

With each YES/NO question, we rule out
half of the squares and gain a single
BIT of information_

How much information was gained by learning the position of the secret
square? The answer is implicit in the technique we used to find the
square: with each answer the information increases by a *bit*. So, after
the sixth question, and having located the square, six bits of information were generated. Generalising this, we can say that a *64-state system*
can encode precisely six bits of information. Of course, we could encode
exactly the same information using six coins, or binary digits, each
encoding a single bit of information. We can extend to this to systems
of any size or complexity, perhaps with trillions of different states.
No matter how many different states a system can occupy — as long as
the number of states isn't infinite — the information it encodes can
always be equivalently encoded using a sequence of *bits*. Later in the
book, we'll discuss how this even applies to your brain, which generates
vast amounts of information by selecting from an incalculable number of
states. In theory, there is no reason why all the information generated
by your brain couldn't be encoded by a long string of binary digits and
uploaded to a computer.

11001010111011001100101011100100111100101110100001101000011010010110111001100111001000000110
11100110010000000110000100100000000110110101100001011011100110101001011011001100110010101110011011110
11000010111010000110100101101111011101110110111000010000000110111101110011011000010000000111010001101000011011
01000000011000110110111101101110011010111000001101100011001010111100001101001011001100110100101101
11000010111010000110100100101101110111011011100001000000011011110111001100110000011010010110111011100110
11011110111001100110110101101100001011110100001101001011011101101110

However, it's estimated that the memory capacity of the human brain is
around 2.5 million gigabytes or 20,000,000,000,000,000 *bits*, whereas
the world's largest supercomputer only has a memory capacity of less
than a third of this.

A system that can exist in one of two distinct states can encode a single bit of information.

In summary, any system that can exist in a finite number of discrete states contains a finite amount of information that can be measured in *bits*. The amount of information encoded by a system can be measured in terms of the number of states that the system can occupy. A system with many states, encoding a large amount of information, can always itself be encoded by a number of simple *2-state single-bit* systems. When a system adopts one state from a number of possible states, information is generated. The amount of information generated depends upon how many states are *ruled out* when one of the states is selected: when I select a single square on a checkerboard, I rule out the other 63 squares and generate 6 *bits* of information. If I use a smaller 4x4 checkerboard, then only 15 squares are ruled when selecting a single square, so less information is generated — 4 *bits* to be precise_

But how does all of this relate to the structure of our Universe? To understand this, we must first break the Universe up into its smallest components and, once we look deep enough, we will find nothing at the ground of reality other than information.

The Ancient Greek philosopher Democritus believed the world, and all within it, to be built from tiny indivisible particles he called ATOMS. Democritus' ideas heralded the birth of the atomic theory of matter: all the world and all within it, barring perhaps the soul, is a complex orchestration, by divine hand or otherwise, of countless pellets of the absolute. The belief that the atom was the fundamental unit of matter reigned until the end of the 19th century, when the ELECTRON — the negatively charged component of the atom — was discovered and, shortly thereafter, in the early 20th century, the atom was successfully split into its constituent electrons and nucleus, which contains positively-charged PROTONS and uncharged NEUTRONS. This was the dawn of subatomic physics and, like the atom itself was shattered, so was the belief that the quest for the fundamental stuff of the Universe was over.

Generations of particle physicists spent the remainder of the century developing what would become the STANDARD MODEL of particle physics. According to the Standard Model, our world is constructed from 17 elementary particles that can be broadly classified as FERMIONS and BOSONS. The fermions can be further classified into two groups: QUARKS and LEPTONS. The electron is a type of lepton, and specific agglomerations of quarks form the more familiar subatomic particles, such as protons and neutrons.

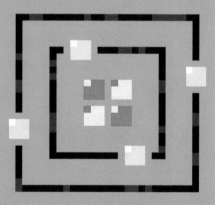

Electrons surround the nucleus in mathematically-defined oribitals_

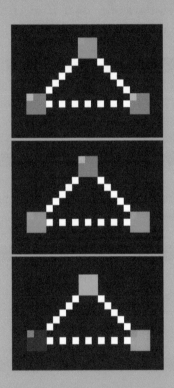

A *proton* is built from three quarks: two UP quarks with electrical charge +2/3 and a DOWN quark of charge -1/3, forming a particle with an overall charge of +1. These three quarks are held together by massless bosons appropriately named GLUONS. Similarly, a single UP quark and two DOWN quarks combine to form the uncharged *neutron*. In general, all non-elementary particles can be defined in terms of compositions of elementary particles and their interactions.

It's tempting to imagine an elementary particle as being simply the smallest piece of matter: if you keep breaking matter up into smaller and smaller pieces, you will eventually reach particles that, no matter how hard you try, cannot be broken up any further. Whilst this isn't entirely incorrect, it assumes that an elementary particle, such as an electron, is somehow a tiny 'chunk' of matter. In fact, an electron (or any other elementary particle) has no *size* or *volume* that would justify thinking of it in this way.

Inside an atom, the negatively-charged electrons surround the positively-charged nucleus (comprising protons and neutrons) in mathematically-defined structures called ORBITALS. Each of these electrons can be defined in its entirety by a set of four numbers called QUANTUM NUMBERS, each of which is *quantised*, meaning it can take only one of a set of discrete values at any point in time. Each quantum number defines a specific property of the electron:

The principal quantum number, N, defines the *energy level* of the electron and can only take integer values from 1 upwards:

```
1, 2, 3, …
```

The azimuthal quantum number, L, defines the angular momentum of the electron and can take integer values from 0 to N-1.

```
0, 1, 2, ... N-1
```

The magnetic quantum number, M(L), defines the direction of this angular momentum and can take integer values between -L and L:

```
-L, ..., 0, ..., L
```

The spin quantum number, M(S), defines the spin angular momentum of the electron, and can only take on two values (often referred to as *spin up* and *spin down*, respectively):

```
+1/2 or -1/2
```

There are no other properties that an electron can possess that can distinguish it from the other electrons within an atom. So, rather than a chunk of matter, an electron is more like a *point* in space tagged with a set of digital (quantised) values. However, since quantum theory doesn't allow us to know the exact position of an electron before it's measured, the quantum numbers only allow us to plot the probability of finding the electron at any point around the atom. The positions with a non-zero probability — where the electron might possibly be located — form a well-defined shape referred to as an *orbital*. Note that the quantum numbers are used to define the orbital, not the other way round.

For example, if an electron's *azimuthal quantum number* is 0, its probability map is spherical: an S-ORBITAL. An azimuthal number of 1 defines a
dumbbell-shaped P-ORBITAL. A fundamental principle of quantum physics is
that only two electrons can occupy a single orbital. Why is this the case?
Simply because adding a third electron to an orbital would require two
electrons to share the same four quantum numbers. And, since an electron
is defined entirely by its quantum numbers, two electrons with the same
set of numbers is simply the same single electron.

All the possible orbitals within an atom are derived from all possible arrangements of the four quantum numbers. And, since each of these
numbers is quantised, the number of possible arrangements is finite.
Equivalently, we can say that *an electron can only occupy a finite set
of discrete exclusive states*, with each state being a particular set of
quantum number values. This applies not only to the electron, but to all
fundamental particles within the standard model — each particle can be
fully defined by a finite set of quantum numbers, each of which can take a
finite set of discrete values. This means that every fundamental particle
within the Standard Model of particle physics can be fully defined by a
finite amount of information, since there is a finite number of possible
arrangements of each particle's quantum numbers.

Using our working definition of information, when an elementary particle is located in a particular orbital, a single arrangement of quantum
numbers — a single *quantum state* — is selected from a finite number of
possible states: this generates the information that defines the particle in the same way that selecting a square on a checkerboard generated
six bits of information. So, a particle is not an object, but only the
information generated by its particular arrangement of quantum numbers.
Theoretical physicist Max Tegmark goes as far as to say:

"The particles [of the Standard Model] are purely
mathematical objects in the sense that they have no
properties at all beyond their quantum numbers."[1]

And since, of course, all matter is composed of these particles, this
means that all matter can be defined by a finite amount of information.
In fact, everything within the Universe is nothing more than quantised —
digital — information.

Each orbital can carry two electrons — one spin-up, one spin-down —
defined by the arrangement of their quantum numbers_

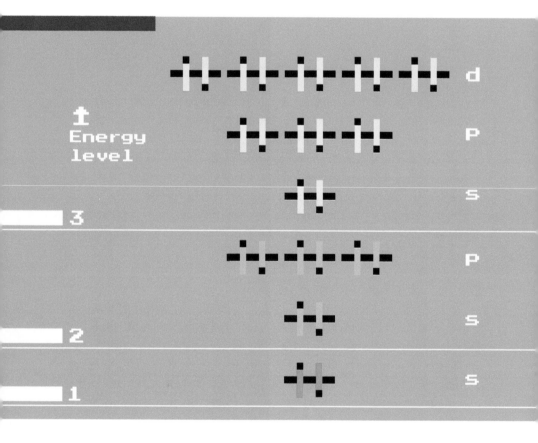

The two electrons in the first energy level are defined by the
following quantum numbers:

1 0 0 1/2

1 0 0 -1/2

Of course, the Universe is much more than a vast set of elementary particles. These particles are dynamic, constantly moving through space,
shifting their energy states, and interacting with other particles to
form larger, more complex structures, including atoms, molecules and,
ultimately, living organisms such as ourselves, in a grand hierarchy of
increasing complexity. However, this doesn't alter their status as being constructed from information: when an electron inside an atom jumps
from one energy level to another, for example, all that's occurring is
a change in the quantum numbers of the electron according to particular
rules.

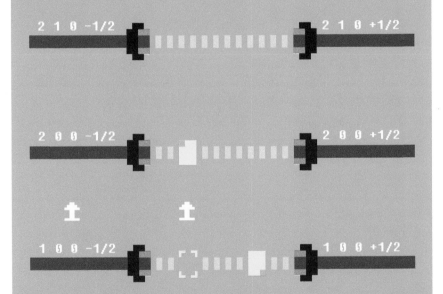

For example, an electron in the lowest energy state — known as its GROUND
STATE — has the principal quantum number, N=1. If the electron absorbs
exactly the right amount of energy from a photon, it can jump to a higher
energy orbital: its principal quantum number increases to 2. *What has actually happened?* The information that defines that electron has changed.
Whilst it's easier to visualise an electron hopping from one orbital to
another, this is merely a convenient image. In fact, all that has occurred
is a COMPUTATION, which we can define as the *processing of information according to rules.* So, the interaction between the photon and the electron
is actually an interaction between two pieces of information. Both the
electron and the photon are defined by information and their interaction
changes this information according to specific rules described as the
laws of quantum physics.

The same principle applies to the merging of several elementary particles to form a larger particle: the interaction between three quarks to form a positively-charged proton, for example. Each quark is defined by its own set of discrete quantum numbers — distinct from those that define electrons — and is thus a finite piece of information. The same is true for the gluons that mediate the interaction that holds the three quarks together.

> So, the information that defines the three quarks and the gluons interacts to form a more complex chunk of information that we refer to as a proton.

In fact, all events that occur at the quantum level — the collision of two particles, the decay of a particle into two or more smaller particles, or the joining of several particles to form another particle — are nothing more than the processing of information according to rules: *computations*. And, by extension, we can describe all processes, at all scales within the Universe, from the microscopic to the cosmic, as part of one huge ongoing computation.

Since the *prima materia* of our Universe is digital information, as with any form of information, it's theoretically possible to encode the Universe in its entirety using a colossal but, crucially, finite number of bits — the best estimate is that around 10 to the 120th power bits would be required. This idea inspired the work of eminent theoretical physicist, John Wheeler, who famously maintained that digital information lay at the ground of reality. His epigrammatic coinage — *"It from bit"* — submitted that

"every particle, every field of force, even the space-time continuum itself derives its function, its meaning, its very existence entirely from the apparatus-elicited answers to yes-or-no questions, binary choices, bits."[2]

It is the processing of these bits according to a rule set that generates the observed physics of our universe. At its deepest level, deeper than the atom and more fundamental than the quark, the Universe is running a low level computation.

Chapter 3:

The Complexi-fication of Information_

In 1969,

inventor of the world's first programmable computer, Konrad Zuse, penned an unusual little book called *Rechnender Raum* —

Calculating Space —

in which he argued that our Universe could be described as a type of computer program that evolves through a series of discrete steps, according to predefined rules. Zuse was proposing that, despite the astounding complexity of the natural world, at its foundation is a set of rather simple rules that compute and update the Universe with each passing moment. *The Universe is computable and is, in fact, computing itself.* In Zuse's model of the Universe, at every point in time, the Universe exists in a specific state which updates with each click of time: every quantum number of every fundamental particle either remains the same or changes according to the rules that govern quantum processes. The Universe updates with one massively parallel computation. This update process also says something important about time. Since the entire Universe updates in parallel, changing from one state to the next, there is no 'in between' state of the Universe, which means time itself is discrete. Time doesn't flow like a river, but is simply the sequence of updates of the Universe from its current state to the next, edging forward one click at a time.

As computer scientist Tommaso Toffoli explains:

"""""In a sense, nature has been continually computing
the "next state" of the universe
        for billions of years;
                all we have to do — and, actually,
                all we can do —
is "hitch a ride" on this huge
ongoing computation"""""[1]

So far, we have discussed how the particles that constitute matter, and time itself, are not continuous, but discrete, occupying a finite number of countable states that encode information. This only leaves space to consider. The emerging field of *digital physics* seeks to explain the physical world in terms of digital information processing, largely inspired by the work of Zuse and Wheeler. One of the founders of this nascent branch of physics is computer scientist Ed Fredkin:

> "If we could look into a tiny region of space with a magic microscope, so that we could see what was there at the very bottom of the scale of length, we would find something like a seething bed of random appearing activity. Space would be divided into cells and at each instant of time each cell would be in one of a few states. A snapshot would reveal patterns of two (or three or four or some other small integer) kinds of distinguishable states.
>
> It would be either pluses and minuses,
> blacks and whites,
> seven shades of gray,
> ups and downs,
> pluses and neutrals and minuses,
> clockwises and anticlockwises or whatever.
> The point is that it would be equivalent to digits."[2]

By assuming that space, as well as time, is quantised – built from indivisible discrete units, which Fredkin calls *cells* – we can begin to build a complete picture of the structure of the Universe as a type of *grid* of cells. Of course, since we live in an apparently 3-dimensional universe, it makes sense to imagine this fundamental grid as being cubic, with each cubic cell surrounded by, or connected to, six neighbouring cells (we could also call this 1:6 connectivity). However, this isn't necessarily the case and other types of grid structures and connectivities are also possible albeit more difficult to envisage.

For ease of visualisation, it makes sense for us to work with a flat 2-dimensional grid for the time being to visualise these cells, but everything we discuss applies equally well to a 3D or higher-dimensional grid.

To understand how these *cells* of space might behave in the type of world envisaged by Fredkin, imagine a tiny region of space constructed from only 16 cells. To keep things simple, each cell can exist in only one of four states (quantum numbers if you like). We will call these states BLACK, BLUE, GREEN, and RED, but we could equally well label them as 0, 1, 2, and 3, or indeed any four distinct names. The BLACK state is the default state and essentially represents empty space. If a cell is in the BLUE state, then we can say it contains a particle we'll call a B-PAR-TICLE. A GREEN state cell contains a G-PARTICLE and a RED state cell an R-PARTICLE.

Just as we've seen with the quantum numbers of electrons, each parti-cle type is entirely defined by its quantum numbers [one in this case]. Similarly, just as there is no 'object' that is an electron, but only information encoded as four quantum numbers, so it is with these simpler particles — it should not be imagined that these states merely represent these particles, but that each particle is nothing more, and nothing less, than the information encoded in a particular state. In the diagram opposite, this slice of space is shown at six time points — 1 to 6 — pro-ceeding anticlockwise from the upper left. At each time step, each cell has updated its state from the previous time step. However, only a small number have updated to a different state, so most cells appear unchanged.

Clearly, the only thing that occurs at each time step is the update of the cell states. However, we can also describe this process from a different perspective: *the G-PARTICLE moves diagonally downwards and to the right until colliding with the B-PARTICLE. Both particles are annihilated and two R-PARTICLES are emitted, which move away perpendicular to each other.* In fact, this process is analogous to what occurs when an electron col-lides with its antimatter partner, a *positron*: they are both annihilated and two photons are emitted. Since both the electron and positron are defined entirely by information encoded in cell states, their collision and annihilation is simply the processing of information — a computation. We can apply this same principle to our own Universe: since everything is defined as digital information encoded as the states of the cells of the fundamental grid of our Universe, there is no requirement for 'objects' to move around in space as such. All that occurs is the updating of cell states according to rules. Information, encoded in cell states, is what moves around the grid.

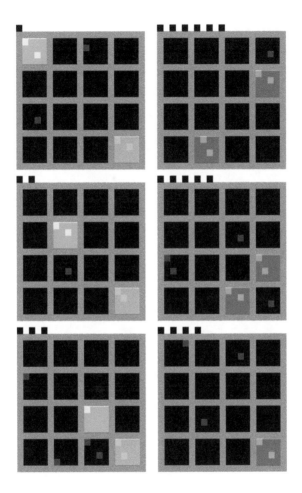

The GREEN particle collides
with the BLUE particle, both
are annhihilated, and two RED
particles are emitted.

Looking around at the marvellous complexity of the natural world, it's hard to imagine that all of this is a manifestation of computation. This is because the effulgence of the manifest world emerges from a hierarchy of increasing complexity, and the fundamental grid is buried deep. At the pinnacle of this hierarchy are, of course, the myriad forms of complex life that fill our oceans, forests, and cities. Although life seems as far from computation as it's possible to stray, life is not a *property* of matter but an *organisation* of matter. Life is a dynamic process, a behaviour, that is dependent on specific informational interactions between molecules and the rules governing them (i.e. computations). The molecules themselves are rule-based organisations of atoms, which are organisations of subatomic particles, and so on down to the cells that constitute the fundamental structure of space.

Although this is conceptually not difficult to understand, it does little to expunge the feeling that complex life is somehow miraculous. And, in a sense, it is. But life is made all the more miraculous by understanding that it all emerges from the processing of digital information. To fully understand this, we need to explore this idea from the bottom up, and to imagine how a complex universe, such as ours, is constructed.

When Konrad Zuse imagined the Universe computing itself, he was thinking about a type of computer program called a CELLULAR AUTOMATON. Although they can take many forms, these programs are most often visualised as a 1- or 2-dimensional square grid. Each square - or CELL - of the grid can exist in a limited number of states. In the simplest type of cellular automaton, each cell is limited to switching between only two states, usually coloured *white* or *black* and often referred to as being *dead* or *alive*. As such, each cell can encode a single *bit* of information.

At each point in time, every cell of the automaton will be either *alive* or *dead*. Of course, this isn't particularly interesting, but the interesting behaviour only emerges as the program runs, which it does by updating the state of every cell simultaneously in a series of discrete time steps [see over]. In addition to the grid of cells with allowable state values, every cellular automaton must also have a RULE SET. These rules dictate how a cell in a particular state must update itself with each 'click' of time. For example, if a particular cell (which we'll refer to as the *central cell*) is *alive*, depending on the state of the surrounding cells (usually, but not necessarily, the direct neighbours, which form a NEIGHBOURHOOD), the rule set will dictate whether the cell remains *alive* (stays black) or becomes *dead* (turns white).

The simplest type of cellular automata, known as ELEMENTARY CELLULAR AU-TOMATA (ECAs), are 1-dimensional automata in which each cell updates its state based on its own state and that of its two immediate neighbours. This means the rule set must specify the next state of a cell for each of the eight possible 3-cell combinations. This also means there are only 256 possible different ECA rule sets, all of which have been extensively studied and each given its own number according to mathematician Stephen Wolfram's binary-based numbering system[3]. Since all the cells of an ECA can be displayed as a single row, it's simple to visualise the evolution of cell states over time by plotting the rows of cells one after the other (in other words, by plotting time in the second dimension).

Rule 110:

Beginning with a random configuration of states, many rule sets exhibit rather uninteresting behaviour: either rapidly converging to an unchanging homogeneous state or to a slightly more interesting, but non-complex, stable or repeating pattern. Some rule sets rapidly descend into apparently chaotic or random patterns, whereas a small number of rule sets display patterns that can be described as COMPLEX. We'll discuss complexity in more detail in the next chapter but, briefly, a complex system is neither highly ordered nor completely random, but somewhere in between.

In an ECA, the next state of a cell depends upon its current state and that of its two immediate neighbours.

Rule 110 (opposite), for example, displays patterns that have a clear structure without being perfectly repetitive. Although certain characteristic patterns emerge and maintain themselves, it's impossible to predict exactly how the automaton will evolve with time. However, even in these most basic of all cellular automata, stable structures — dubbed *particles* — that propagate through time can be observed moving through the automaton, decaying to form other particles, or colliding with each other in patterns eerily reminiscent of those generated by collisions inside particle accelerators.

Time

↓

Evolution of Rule 110

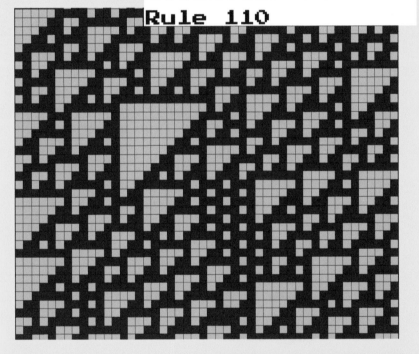

Rule 110 exhibits complex behaviour, which is neither random nor banally repetitive.

Whilst ECAs demonstrate how extremely simple computational rules and an update algorithm can give rise to surprisingly complex patterns, it's hard to relate their behaviour to the sort of complexity we see in the world around us. A closer approximation is provided by higher-dimensional automata, in particular by the most famous of all cellular automata, the *Game of Life*[4]. This 2-dimensional cellular automaton was discovered by mathematician John Conway and has a number of properties that make it particularly interesting.

As with a 1D ECA, the state of each cell in a 2D automaton updates based upon the states of the cells in its *neighbourhood*. Whilst the neighbourhood in a 1D ECA always comprises the two immediate neighbours, there is no reason, in principle, why these particular cells must be chosen. For example, rather than just the single immediate neighbours, one might choose the first two cells on each side of a central cell or, in fact, any set of cells one desires, just as long as they are properly specified in the rule set — that is, for every possible arrangement of neighbourhood cell states, there is a rule that tells the central cell how to update depending on its own current state. However, as the number of cells in the neighbourhood increases, the number of possible rule sets increases exponentially, making exploration of all different automata more and more computationally expensive.

There are two neighbourhoods most commonly employed in 2D cellular automata: the MOORE NEIGHBOURHOOD comprises all eight cells surrounding a central cell, whereas the VON NEUMANN NEIGHBOURHOOD only contains the four cells directly contacting the faces of the central cell:

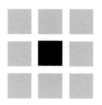

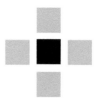

The Moore
neighbourhood

The Von Neumann
neighbourhood

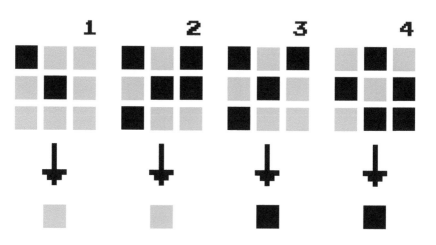

1

If a cell is alive (black) and less than two of its neighbours are alive, it dies of loneliness (becomes blue).

2

If a cell is alive and more than three of its neighbours are alive, it dies of overcrowding (becomes blue).

3

If a cell is alive and two or three of its neighbours are alive, it stays alive (remains black).

4

If a cell is dead (blue) and exactly three of its neighbours are alive, it becomes alive (becomes black), otherwise it stays dead.

The *Game of Life* rule set might seem somewhat arbitrary, and in a sense it is: these rules weren't designed so much as discovered. As with the 1D ECAs, there is only a limited, but much larger, number of possible rule sets, which we will refer to as the RULE SPACE. However, there is no simple way to know whether any particular rule set will display interesting behaviour once the program runs — the only way to find out is to run it. The rule set of the Game of Life was discovered by Conway during a process of simple trial and error. Again, as with the elementary 1D automata, many rule sets show completely trivial and boring behaviour: no matter which cells are alive at the beginning, they all soon die off once the program starts running. At the opposite extreme, the cells descend into chaos, and no interesting forms are discernible. However, the Game of Life rule set is different:

> as the program clicks along,
> the cells eventually begin to form structures that resemble
> a primitive form of artificial life.

A huge variety of structures have been observed, and continue to be discovered. These range in complexity and dynamics, from completely static *still lifes* to stationary pulsating *oscillators* to *spaceships* that glide across the grid emitting projectiles. These projectiles often interact with other structures in surprising ways, causing them to move away from the spaceship or to break apart into new structures.

This oscillator cycles between three symmetric states

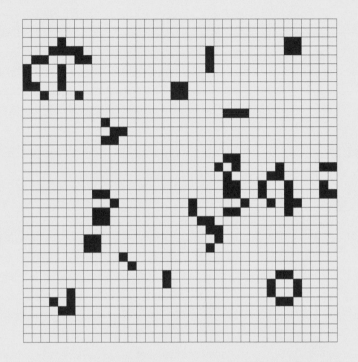

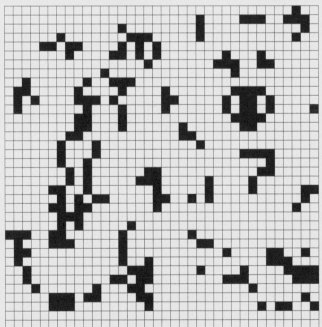

A Game of Life grid at two different time points_

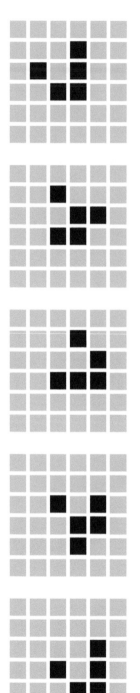

For want of a better term, we will call these structures CRITTERS (in complex networks of simple interacting components, these are often known as AGENTS). One of the most well-studied of these critters is the GLIDER:

A glider is a 5-cell structure that traverses the grid by cycling through four distinct cell configurations.

Although the glider clearly maintains a structure that allows us to recognise it as such, its constituent cells change with each time step. As such, it is somewhat misleading to think of the glider as an object. It makes more sense to think of the glider as a process, or a behaviour of the underlying grid, in the same way that an ocean wave is a behaviour of water. The glider is a sort of wave of state changes rippling across the grid. This also applies to any other type of critter manifested on the grid.

The glider cycles through four 5-cell configurations as it glides across the grid.

Each cell of the Game of Life is a simple type of computer called a finite state machine. Since its constituent cells change with each time step, a glider is a virtual state machine (VSM).

Before we can think about the relationship between cellular automata and the structure of our reality, we need to look at these critters more closely. If we think of the entire grid as a sort of 2D universe, then the rule set that dictates how the cells update with each time step represents the 'laws of physics' obeyed by this universe. The entire universe is itself a computer, since it exists in a specific state at each time point and uses the rule set to compute its next state. However, each cell on the grid is also a simple type of computer known as a FINITE STATE MACHINE (FSM). As the name suggests, an FSM can only occupy one of a finite number of states at any point in time, and can receive a finite number of inputs that it uses to compute its next state (the output of the computation). A cell on an automaton grid can only exist in one of a finite number of states — two in the case of the Game of Life — and receives information from cells in its neighbourhood — its input — to compute its next state, which is its only output.

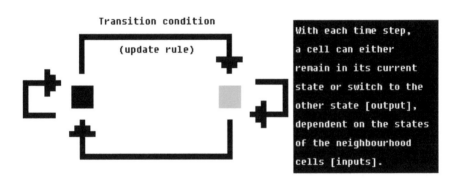

Transition condition
(update rule)

With each time step, a cell can either remain in its current state or switch to the other state [output], dependent on the states of the neighbourhood cells [inputs].

A more complex critter constructed from a group of cells is also a finite state machine: the states of its constituent cells constitute its current state. The states of the cells in the neighbourhood of each critter cell act as the input, with the next state being computed based on this input and according to the rule set. Critters that propagate around the grid, such as gliders, are a little different: since their constituent cells change over time, they are never constructed from the same cells for more than a single time step. As such, moving structures are more correctly known as VIRTUAL STATE MACHINES (VSMs)[5]. So, we no longer view the cellular automaton grid as a single entity, but as a sort of ecosystem of VSMs embedded in the grid. However, at the same time, this entire logical universe is unified by the grid. In fact, it simply *is* the grid.

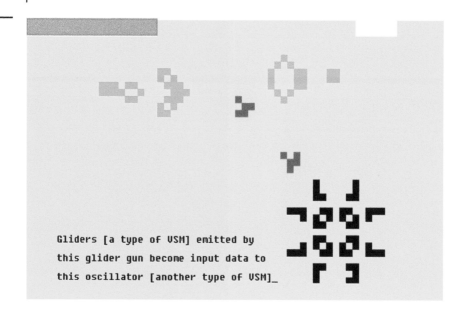

Gliders [a type of VSM] emitted by this glider gun become input data to this oscillator [another type of VSM]_

As a critter — VSM — approaches another critter, each will act as input data to the other, potentially changing the way each critter behaves. Similarly, a critter that emits a projectile can be viewed as sending data across the grid that can be received by another critter. You should notice that the distinctions between structure, process, and data are beginning to blur. If we take the glider as an example: this particular propagating critter occupies five cells at any time point, but cycles through four different configuration states as it moves across the grid. As this glider approaches another critter and enters its neighbourhood, it acts as data that the 'receiving' critter will use to compute its next state update. This data may cause the receiving critter to change its behaviour, its structure, or even to disintegrate completely.

So, this gives us yet another way of thinking about propagating critters: as signals carrying information. Whether we think of these critters as

structures,

processes,

or signals,

depends only upon our perspective. Ultimately, they are ways of thinking about the same underlying digital computations governed entirely by the allowable states of the individual cells and the update rule set.

Things become even more interesting when we consider hierarchies of VSMs. The glider is a configuration of the basic cell finite state machine, and so what we might call a FIRST-ORDER VSM. However, we needn't stop there, since configurations can occur at higher order scales. Multiple first-order VSMs can be embedded in higher order structures: SECOND-ORDER VSMs, and so on. As these high-order VSMs may be comprised of many lower-order VSMs, they become increasingly complex and are capable of performing more and more sophisticated and elaborate computations. The perfectly homogeneous 2D array of identical FSMs — cells — constitutes the 'ground of reality' of the cellular automaton universe, and the highly heterogeneous population, or society, of complex, multifaceted, and sophisticated VSMs is a manifestation of this basic underlying parallel computation.

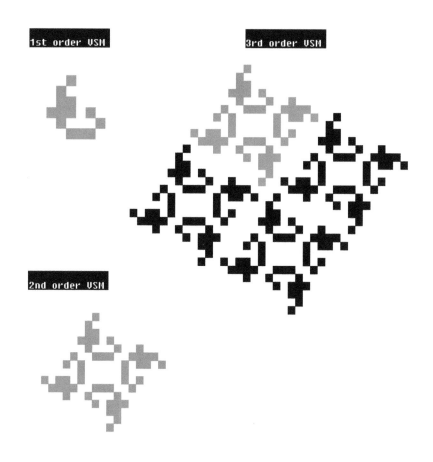

1st order VSM

3rd order VSM

2nd order VSM

Hopefully, this idea of hierarchical VSMs will resonate with the ideas discussed earlier in this chapter:

at its deepest level, the Universe is quantised
and constructed from digital information,
computing its next state according to a
fundamental rule set
as time
clicks along.

If we render the Universe as a 3-dimensional cellular automaton – which we will call the GRID – it's simple to imagine how a high-order structure, such as an atom, can manifest.

Instead of a single layer of 2D cells, we now have cubic CELLS extending in all three directions – a 3D grid of the sort envisaged by Fredkin. *We will use the capitalised Grid and Cell to indicate that we're talking about our Universe cellular automaton and its constituent cells.* Just like the 2D Game of Life, each of these Cells can exist in a limited number of states which update in parallel with each time step. And, just as the 2D automaton can form stable structures on the grid, so it is with the Grid of our Universe. The so-called fundamental particles that manifest in our Universe are the 3D equivalents of the low-order VSMs that manifest on a 2D cellular automaton grid: patterns of Cell states. And the atoms built from these fundamental particles are the 3D equivalents of the higher-order VSMs built from lower-order VSMs on a 2D automaton. From here, it is conceptually simple to see how more complex structures, such as molecules and, ultimately, living cells and organisms, are 3D VSMs even higher in the hierarchy, built from lower order VSMs.

On the 2D grid, we saw that VSMs propagating across the grid change their constituent cells with each time step – better described as dynamic processes than structures. This also applies in our higher-dimensional Universe:

an electron moving through space is a stable process – pattern of Cell states – but is built anew as it traverses the 3D Grid. The same, of course, applies to even higher-order structures, including ourselves. Philosopher Alan Watts expressed this idea using the familiar example of a whirlpool:

"The whirlpool is a definite form, but at no time does water stay put in it. The whirlpool is something the stream is doing, just as we are things the whole Universe is doing."[5]

The glider is something the 2D automaton is doing, and you are something the Grid is doing.

Since every cell of a 2-state cellular automaton can occupy one of only two states, each cell encodes a single bit of information. A number of these cells can naturally form structures that encode more information, such as a glider. We will call these structures INFORMATION COMPLEXES, since they are nothing more than patterns of information. In our Universe, this is also the case: the fundamental Grid instantiates the information that manifests as structures of increasing complexity. This gives us a clear picture of the layered nature of reality built from information complexes in a hierarchy of increasing complexity from the Grid upwards.

Reality is nothing more, and nothing less, than manifest patterns of information — *information complexes* — emergent on the fundamental Grid of the Universe. This includes you, your brain, and your entire world. Of course, you feel like much more than a high-order virtual state machine on a 3D automaton Grid. *You are alive, you are conscious*. We'll get to the conscious part later, but first we must deal with life.

The Hierarchy of Complexity

The Code:
the fundamental code that generates the Grid (discussed later).

The Grid:
the fundamental structure of space consisting of Cells in specific states.

Fundamental particles:
electrons, quarks, neutrinos, etc. Low order information complexes self-organised from the Cells of the Grid.

Subatomic particles:
neutrons, protons. Higher order information complexes self-organised from fundamental particles.

Atoms:
carbon, oxygen, nitrogen, etc. Higher order information complexes self-organised from subatomic and fundamental particles.

Molecules:
proteins, water, carbohydrates, nucleic acids, etc. Very high order information complexes self-organised from atoms.

Cells:
smallest component of life. Higher order information complexes self-organised from molecules.

Multicellular organisms:
higher order information complexes self-organised from cells.

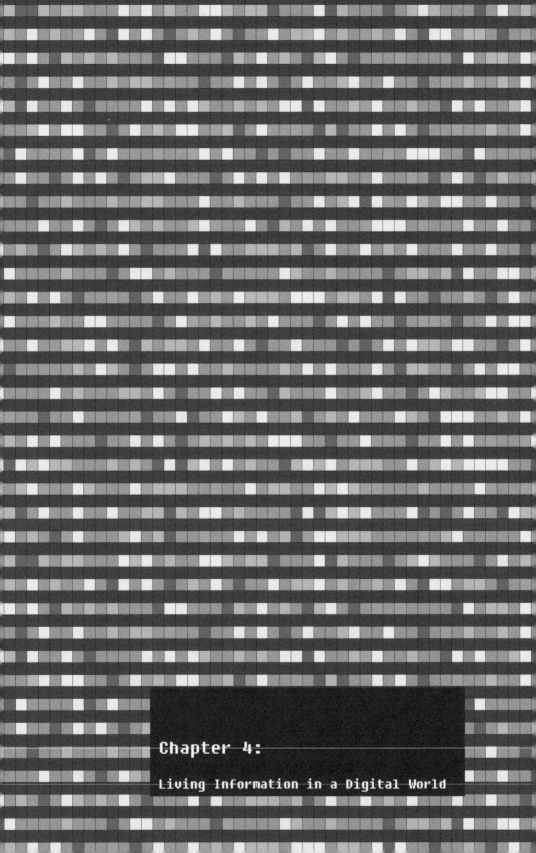

Chapter 4:

Living Information in a Digital World

You can open your eyes and see all this world emerging

out of

nothing...

BOING!!

Alan Watts

Donning a virtual reality headset for the first time, it's startling how the pixellated digital landscapes of early computer games have been superseded by entire worlds within which to wander and become lost. It's almost impossible to fathom how these immersive worlds of such dynamic and responsive complexity and beauty are built from binary code. *The code is hidden, but the code is all.* And, although the immeasurably rich diversity of complex organisms that crawl, slither, and swim around our world seem to be animated by some magically vital principle beyond the physical, each and every one emerges in a hierarchy of complexity from the fundamental Grid at the ground of our reality. To understand the structure of this reality, we need to examine with some care the structure of life. For although life feels somewhat disjoint from the cold, dark, and dead universe that surrounds us, even your own undeniable subjective reality is built from digital information. *Everything is a manifestation of the complexification of information.*

Imagine a society of social insects, such as an ant colony. A very large colony might contain several million ants, with different ant types providing specific roles in the organisation of the society: the *queen* is the leader and founder of the colony, her eggs fertilised by the male *drone* ants. Female *worker* ants are responsible for maintaining the nest, as well as protecting it from intruders and predators. Each individual ant is a rather simple creature and interacts with the other ants in rather simple ways. However, the entire colony can display highly sophisticated behaviours: building elaborate networks of underground tunnels, delineating routes to food sources, defending against coordinated external attacks, or even forming themselves into bridge-like structures over water. No single ant directs the behaviour of the colony and no single ant, not even the queen herself, is aware of the overall structure or purpose of the colony. The colony SELF-ORGANISES and its behaviour emerges from the myriad simple interactions of its members.

The complex behaviours displayed by the ant society only emerge when the society is functioning as a whole. These are EMERGENT BEHAVIOURS, which are a characteristic property of so-called COMPLEX SYSTEMS and can be defined as *behaviours or properties exhibited by the entire system but not by its parts*. The whole is greater than the sum of its parts. Each individual ant behaves in a simple, stereotypical manner, but the entire colony exhibits sophisticated behaviours above and beyond that of any single ant.

Although perhaps hitherto unnoticed, emergent behaviour is everywhere. Anyone who's seen a flock of starlings – a *murmuration* – sweeping, twisting, and reconfiguring in elaborate dynamic patterns across an autumn sky can hardly fail to be astounded by its remarkably intelligent-looking behaviour. However, the flock is not directed by any particular starling and it's likely that individual birds mainly use their closest neighbours – *local interactions* – to direct their movement in the air. *Flocking* behaviour emerges not from the individual birds, but from the interactions between them. Just as there is no 'society programme' encoded in every single ant within a colony, flocking is not programmed into the individual birds. The organisation of birds into dynamic flocks, or ants into complex societies, emerges from the multitude of rule-based interactions between individual birds or ants. The flock doesn't need to be actively organised by a central organiser – the flock self-organises.

From these two examples alone, we can begin to delineate the essential properties of what are known as COMPLEX SYSTEMS. A complex system is not just a complicated, chaotic or random system, but one with at least four basic characteristics:

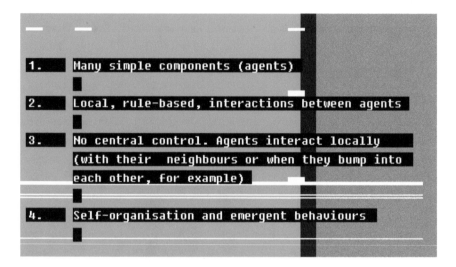

1. Many simple components (agents)

2. Local, rule-based, interactions between agents

3. No central control. Agents interact locally (with their neighbours or when they bump into each other, for example)

4. Self-organisation and emergent behaviours

In the case of a flock of starlings, it's easy to spot these four features: the flock is constructed from many simple components: the individual starlings. These interact by observing and responding to the position and flight of neighbouring birds — local interactions — and, whilst there is no central bird controlling the flock, the birds self-organise and the beautifully dynamic behaviour known as *flocking* emerges from this self-organisation. Of course, not all types of birds demonstrate this flocking behaviour: Self-organisation is organisation without an organiser and, if the rules are right, order comes for free. But the rules must be right. Not all systems with many interacting components will exhibit complex emergent behaviour. To understand why this is the case, we will return to our exemplar cellular automaton — the Game of Life.

All cellular automata display the first three characteristics of complex systems: they are constructed from a large number of simple components — these are the fundamental cells that form the grid. Each of these cells interacts locally with the cells in their immediate neighbourhood according to a rule set. And, like a flock of starlings, there is no central cell that controls the behaviour of the automaton — all cellular automata are completely democratic.

However, despite the obvious commonalities in their construction, not all cellular automata display what could be called complex behaviour. The patterns generated by an automaton as it runs depend upon its rule set, and the behaviour it displays will always fall somewhere along a continuum of complexity: at one end is perfect crystalline order and, at the opposite is the entirely unpredictable, thoroughly disorderly behaviour known as *chaos*. Mathematician Stephen Wolfram exhaustively studied the rule sets of many types of cellular automata and observed their behaviour, noting that different rule sets often produce startlingly different behaviour, but which falls somewhere on this scale. Wolfram defined four distinct types of behaviour:

Class I: A fixed, unchanging pattern eventually emerges.
Class II: A pattern consisting of periodically repetitive regions is produced.
Class III: A chaotic, apparently random, aperiodic pattern is produced.
Class IV: Complex, localised structures are produced.

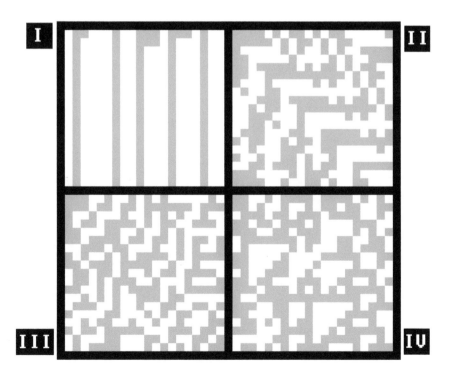

We met these categories less formally in the last chapter when discussing the different behaviours of elementary cellular automata, but let's look at them more closely. *Class I* cellular automata may display early flurries of activity, but this soon settles into an unchanging pattern with all cells remaining in the same state. *Class II* automata are characterised by repeating patterns that are slightly more interesting but highly ordered and non-complex, whereas *Class III* automata display a tumult of chaotic behaviour that never settles. Complexity, the domain of the *class IV* automata, is neither order nor chaos, but sits in a narrow band between *class II* and *class III* behaviour known as the *edge of chaos*. This special region is where complexity reaches its maximum before descending into chaos, and it is only within this narrow band of the complexity continuum that stable propagating structures, such as *gliders* and *spaceships*, can emerge and remain stable. A complex system maintains a type of order, but it is not a static order unresponsive to the environment — it is a dynamic order. A flock of starlings holds together as an ordered structure which we can recognise as a flock, whilst remaining fluid, dynamic, and ever-changing. This is the hallmark of a complex system at the *edge of chaos*, where order and disorder are critically balanced.

The Game of Life exemplifies this type of complex behaviour: as the game runs, the grid begins fizzing with a diverse range of critters — VSMs — that propagate around the grid and interact with each other. Some of these *1st-order VSMs* may join together to form higher-order VSMs and, eventually, the grid is transformed into a dynamic 'ecosystem' of VSMs at varying levels of complexity. Although each individual cell of the grid is relatively unsophisticated, the emergent behaviour of the system as a whole is exquisitely complex and unpredictable — simple rules can give rise to highly complex behaviour. However, although the patterns emergent on a Game of Life grid are certainly fascinating — even life-like — it's not difficult to appreciate that they emerge from the rule-based computations of the individual cells of the grid. As such, the Game of Life offers us a clear example of the *complexification of information*: The critters that emerge on the grid are built from information instantiated by the states of individual cells and, as these critters appear to move about the grid, it is actually *information* that is moving. This flow of information across the grid, instantiated by the rippling of gliders or other structures is critically important for generating the complex behaviour that distinguishes the Game of Life from non-complex automata.

The critters that emerge on the Game of Life grid fall into a number of categories, with the simplest being the static structures known as *still lifes*. Whilst these might be interesting to a limited extent, a grid populated entirely by still-lifes could never be described as complex, since still-lifes have no way of interacting with each other, ever remaining islands irrevocably separated. Rather, it is the dynamic, propagating critters that allow complexity to emerge. As these *gliders*, *spaceships*, and other dynamic critters scuttle across the grid, they can interact directly, by colliding with each other, or indirectly, by emitting projectiles that interact with other critters in their path. As in all complex systems, these *interactions* between the critters – the *agents* of the system – engender the emergent complexity. In the case of the Game of Life, it is the interactions between moving *patterns of information – information complexes* – that generates the complex behaviour.

Whether a particular cellular automaton exhibits complex behaviour will depend to a large extent on whether or not it can generate such stable but dynamic patterns that interact and transmit information around the grid. A *class I* automaton, which remains unchanging over time, is obviously unable to generate such patterns. A chaotic *class III* automaton is equally unable to transmit information: any structures that form are transient, quickly disintegrating as the information they encode dissipates into the seething sea of chaos. But, in *class IV* automata, at the *edge of chaos*, stable structures can emerge and maintain their form as they glide across the grid, carrying the information they both encode and, indeed, are built from. So, *class IV* automata can be distinguished from their less interesting *class I-III* cousins by the way information flows through them, allowing stable patterns not only to emerge but to interact.

The stable propagating and interacting VSMs that characterise the Game of Life form a complex system at a level *above* the base of the grid: The individual cells are, of course, unmoving with a static and unchanging set of neighbours with whom they can interact. In this sense, all cellular automata are completely static — only the states of the individual cells change over time, with each parallel update of the automaton. It is only the information instantiated in the cell states, carried by the VSMs, that flows about the grid. The VSMs form a system of interacting agents at an organisational level above, but emergent from, the fundamental static agents — the cells — at the base of the grid. And, this can occur at many levels of an organisational hiearchy. Information complexifies *upwards*, with new complex systems of interacting patterns of information emerging at higher and higher levels of an organisational hierarchy, the formation of each driven by the flow of information between these patterns and through the system. In non-complex — *class I-III* — cellular automata, this complexification never progresses above the level of the base grid.

This same principle applies in our universe, in even the most complex of structures, including conscious living organisms. No objects really move in the Universe: patterns of information emerge, instantiated by the Cell states of the Grid, and this information flows between Cells generating the patterns of information that self-organise to form the high-order patterns we recognise as atoms, molecules, cells, living organisms, and the patterns *they* form, whether it be a flock of starlings, a society of ants, or human civilisation. The layered complexification of information instantiated by the Grid generates all the forms we observe in our Universe. The difference between our Universe and the Game of Life grid is that this layered complexification has progressed much higher in the Universe, with a deep organisational hierarchy emerging, with living organisms sitting at the apex.

All complex systems can be recognised as self-organised emergent patterns of information. What we recognise as 'objects', together with the names we use to label them, are merely descriptive conveniences that allow us to describe and distinguish all the different patterns of information at varying levels of complexity that are emergent on our Universe Grid.

So what about life?

It shouldn't come as any surprise that living organisms are rightly cat-
egorised as complex systems and, as such, are most accurately viewed as
highly complex patterns of information processing.

Life is an emergent behaviour of certain complex systems.

Living things must, of course, maintain a degree of order — otherwise they couldn't exhibit the complex, meticulously-organised functions
characteristic of living organisms. However, this order must be balanced
by a certain degree of flexibility, since a rigidly ordered system would
be unable to respond to information from its environment, process information in novel ways, make decisions, or evolve over time.

*Too little order and life breaks down, but too much and life
cannot form.*

As with other complex systems, life can only exist in that narrow band at
the *edge of chaos* where order and disorder are critically balanced. But
life seems like a very special type of complex system and, indeed, life
has particular properties that allow us to define it as such.

The patterns of information referred to as living organisms sit high
within a stack of layers of increasing complexity. At every layer of a
hierarchically-organised complex system we can observe emergent behaviours not exhibited by lower levels, with emergent behaviours at one level allowing new, even more complex, behaviours to emerge at even higher
levels. Life represents the pinnacle of this process of self-organisation
and the complexification of information, taking place on the Grid of our
Universe. Life is a special type of emergent behaviour that only appears
at the highest levels of an organisational hierarchy, built upon many
layers of diverse emergent behaviours below it. *But what exactly is life?*

Most people would feel fairly confident in distinguishing something
living from something non-living: Birds, spiders, plants, and bacteria
appear obviously alive, but pinning down a precise definition of 'life'
is more tricky. Perhaps the pithiest definition came from biologists
Humberto Maturana and Francisco Varela, who posited that living organisms are distinct from non-living machines in that they are AUTOPOIETIC,
meaning 'self-creating'[1].

High-order,
emergent
structures

Isolated
simple agents

...nplexity

Although a glider on a 2D cellular automaton is certainly not alive, the glider maintains its identity despite its constituent cells changing every few time steps. Similarly, living organisms are not objects as such, but patterns of information that maintain and regenerate themselves over a period of time we recognise as a lifespan, which might be a few days or several decades.

The simplest organism recognised as living is a single-celled organism, such as an amoeba. Although, throughout its short life, every single component of the amoeba might be degraded and replaced several times, it still maintains its identity as an amoeba. It achieves this by virtue of its construction, from complex networks of interacting components that maintain themselves despite individual components breaking down. As old components of networks degrade, other networks act to replace these components. So, the entire system regulates, maintains, and regenerates itself. Put simply, in the words of physicist Fritjof Capra:

> *"Life is a factory that*
> *makes itself from within."*[2]

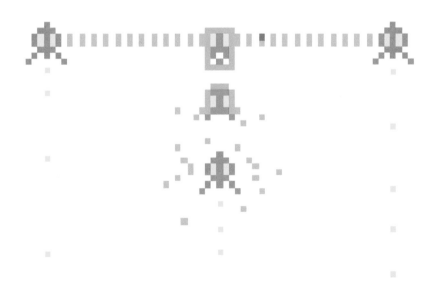

The process of self-maintenance, regulation, and regeneration, is an emergent behaviour, a highly complex type of information processing. This behaviour emerges from the interactions of large numbers of simple components — so a living organism is not a large complex machine so much as a system of smaller and simpler machines, which are themselves patterns of information. It is the interaction of these small machines that produces the complex behaviour of living organisms and, fundamentally, these interactions are the flow of information through the system. You should be able to see the relationship between the emergent behaviour of a flock of starlings, or a colony of ants, and a living cell, which one might go so far as to call a 'flock of molecules'. In general, *a living organism is a complex, hierarchically-organised pattern of information that maintains, regenerates, and replicates itself over time.*

If a cell were large enough for you to open it up and peer inside, you would find it packed with small molecules swimming in the gel-like interior, interacting with other molecules and building new ones. Like all complex systems, cells are built from a large number of simple components — proteins, sugars, lipids, DNA, etc — that interact with each other. Whilst the individual interactions between components are relatively simple, it is from these interactions that highly complex behaviour emerges: building and maintaining the structure of the cell, absorbing and processing nutrients from the environment, excreting waste products, moving away from toxins and towards food sources, and even splitting to form new cells. The complex behaviours of the entire cell emerge from the multitude of interactions of the simple components, and it is these emergent behaviours that, together, produce an *autopoietic system* that we consider to be alive. Of course, in multicellular organisms, such as ourselves, large numbers of cells can self-organise to form even more complex structures, such as muscle, hearts, and brains. In the human body, we can see a number of higher levels of organisation:

cells self-organise to form tissues;
tissues self-organise to form organs;
    and organs self-organise to form the complete organism,
    with new
        emergent properties emerging at
        each layer of the hierarchy.

Whilst a simple cellular automaton such as the Game of Life could never be considered alive, the behaviour of these *class IV* automata is useful in revealing how simple rules can give rise to complex, hierarchically-organised, patterns of information. Despite having only two cell states — *dead* or *alive* — the Game of Life displays a striking level of complexity, with a multitude of critters emerging and scrambling across the grid. As they interact, these first-order critters — VSMs — can self-organise to form high-order structures with emergent computational capacity, some of which begin to emit — construct — other critters. These new critters become part of a diversifying ecosystem, themselves interacting with other critters to form novel high-order critters. Eventually, the grid begins frothing with an increasingly complex and diverse array of critters, interacting with each other in increasingly complex ways. It is not too much of a stretch to imagine how autopoietic — and thus living — structures might eventually emerge from this sort of rich 'biochemical' soup.

The Game of Life will always remain on the first few rungs of the ladder of complexity, whilst our Universe has progressed much higher and, most likely, will continue to do so. Despite sitting at very different levels of the organisational hierarchy, both our Universe and the critters in the Game of Life emerge from information processing. In the case of the Game of Life, it is the cells of the automaton grid that perform the computations, whereas in our Universe, it is the Cells of the Grid.

This picture of living organisms as autopoietic systems of hierarchically self-organised patterns of information is comprehensive but not yet complete. Living organisms don't exist in a vacuum but, rather, embedded in an environment. In fact, they are both built from the same stuff as the environment and fully dependent upon it. For an organism to remain alive, it must depend upon a flow of molecules and energy into and out of its cells, which we recognise as food, oxygen, and waste products. From the informational perspective, this is nothing more than the flow of information from the environment into the network of interacting patterns of information inside the organism.

For example, embedded in the membrane of all cells are specialised proteins called *receptors*, which are charged with transmitting information across the membrane, from the environment into the network of proteins and other molecules inside the cell.

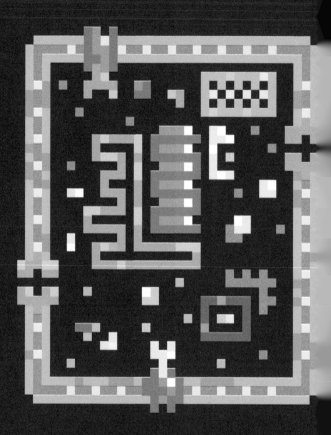

A cell is a self-organised emergent structure, built from a large number of simple interacting components, each of which is itself a self-organised structure with emergent properties_

A receptor spans the entire membrane, with an outward-facing domain exposed to the environment, and an inward-facing domain exposed to the cell's interior. The outer domain binds selectively to molecules in the environment, with different receptors binding to different molecules. Upon binding to the receptor, the molecule causes the receptor to twist out of shape, distorting both its outer and inner domains. The altered geometry of the inner domain allows it to bind to molecules inside the cell known as *signalling proteins*, which are themselves part of the complex network of molecules inside the cell. So, overall, by binding to the receptor, a molecule outside the cell transmits information across the membrane and into the cell. By sporting a variety of different receptors in its membrane, each binding to a specific type of molecule in the external environment, a cell can receive a variety of forms of information about the environment. It's not difficult to see why this could be useful: it might, for example, allow a cell to detect sources of food or toxins, chemical signals secreted by predators, or even to distinguish between light and darkness.

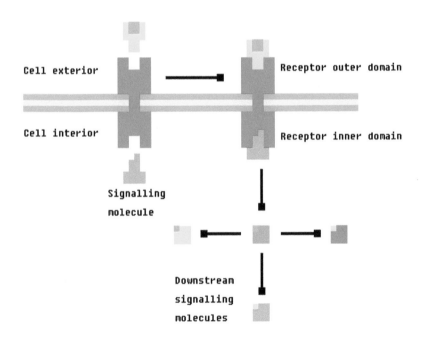

Cell exterior   Receptor outer domain

Cell interior   Receptor inner domain

Signalling
molecule

Downstream
signalling
molecules

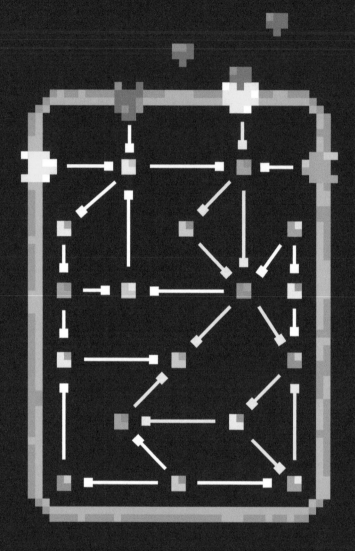

When a molecule binds to a receptor protein,
information is transmitted from the environment
into the cell, where it spreads through, and is
processed by, the network of molecules in the
cell's interior_

The binding of a molecule to a receptor protein exemplifies one of the most important concepts in this book: *perception*. In humans, perception is commonly defined as the process by which awareness of the world is achieved via the senses: the detection of light by the eyes, of sound vibrations by the ear drum, for example. Although we will eventually discuss how your brain is able to perceive orthogonal dimensions of reality during a DMT trip, we must first think about perception more generally and on a more fundamental level. Perception is usually described as a means of sensing the environment and is typically restricted to organisms that are either presumably conscious or that, at least, possess a nervous system. However, it will be helpful to define perception more generally as *the process by which a structure, living or otherwise, receives and processes information from its environment.*

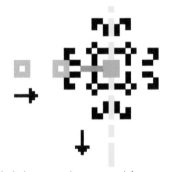

A receptor embedded in a membrane enables a cell to select information from the environment and to transmit this information across the cell boundary and into the network of molecules inside the cell. This is perhaps the most basic form of perception of which all organisms must be capable. An organism will not survive for long if it cannot receive, and process, information about its environment, which is the basis of perception. A system of receptors embedded in a cell membrane, each specialised to receive and transmit a specific type of information, allows an organism to thrive within an ecosystem of other organisms competing for food and other resources. For example, receptors that bind glucose molecules can provide information about the location and concentration of this important energy resource. Similarly, receptors for toxins, pheromones, or other significant molecules, provide an organism with a kind of chemical map of the environment, with concentration gradients of specific molecules mapped out by their differential activation of receptors on different parts of the cell membrane.

Of course, it isn't necessary for the molecules themselves to pass across the cell membrane, only that they bind to the external domain of the receptor such that the information signalling their presence is transmitted into the cell. This system of receptors allows the cell to continuously receive information from the environment, which can then be processed by the network of molecules inside the cell. Naturally, this information is only useful if the cell has the machinery to 'make sense' of it and respond in a functionally useful manner. This requires not only perception, which we can also call *sensory input*, but also a means of producing a response, which is the *output*. This response might be rather simple and automatic, such as moving up a concentration gradient towards a food source or away from an area with a high concentration of toxins. However, whilst such stereotypical responses might be sufficient for a single-celled organism, a multicellular organism perched several layers in organisational complexity above such amoebic simplicity, requiring of a constant and intensive supply of nutrition, and with the constant threat of being preyed upon by other complex multicellular organisms, this would never do.

The more complex an organism becomes, the greater is its need for a rich real-time supply of information from its environment. And, not only must this supply be abundant but, if the response in any given situation is to be appropriate and effective, this information must also be filtered and processed. As an organism evolves towards greater and greater complexity, a specialised structure with the sole purpose of receiving and processing information from the environment becomes a necessity. Of course, this is what we know as a brain*, the biological information generator and processor *par excellence*. As brains evolve and complexify, they begin to construct elaborate models of the environment, containing a wealth of information pertinent to the organism's survival. Eventually, an information-rich subjective map that constitutes the personal reality of an individual organism emerges: *the organism wakes up in a world.*

*we will often refer to the structure that performs the function of a brain more generally as a *brain complex* (a type of information complex), since such structures are universal features of complex life, whether Earthly or otherwise.

Chapter 5: Waking Up in a World

Your brain is the most complex structure in the known Universe, effortlessly manifesting the rich and lucent tapestry of your world from moment to moment. As you witness this glorious unfolding of the world from behind your eyes, you'd be forgiven for losing sight of its fundamental nature: as information generated by your brain, which is itself an information complex instantiated on the Grid.

Each of us lives out each of our lives from behind our eyes, in a world that French philosopher Pierre Teilhard-de-Chardin called the 'within of things': "the object of a direct intuition and the substance of all knowledge"[1]. It's not enough to say that your body and brain is built from information: For conscious beings such as ourselves, borrowing from philosopher Thomas Nagel, there is something-it's-like-to-be this particular configuration of information. There is a within, and this within is your phenomenal world, the world of subjective experience and the only world you can ever know. Whether you are hurrying down a damp street on a grey Sunday morning or dancing with ancient gods in a jewelled temple with bizarre crystalline geometries of immeasurable luminosity and form, it is always your own personal world, and it is always built from information generated by your brain.

If we are to understand the bizarre worlds visited at the peak of a DMT trip, we must first understand the more prosaic world of your daily life in this universe: its purpose, why it takes the form it does, and its relationship to everything happening out there in the so-called external world.

Transcendental idealist Immanuel Kant distinguished between the PHENOMENON — the subjective world within which each of us lives — and the NOUMENON — the unknowable world-in-itself, independent of our perceptions. Your brain is a high-level information complex that generates the information that is your phenomenal world — this is the phenomenon. The noumenon is all the information being generated by the Grid:

the world-in-itself is information.

Your brain generates a highly complex form of information that has the special property of subjectivity: you experience this information as your world, whether you are awake, dreaming, or at the peak of a DMT trip.

Of course, although your phenomenal world is always built from information generated by your brain [we will discuss later how your brain
achieves this], there is an undeniable relationship between this subjec-
tively-experienced world and the world outside of your brain, which we
would naturally call the *environment*. Your brain evolved as a generator,
receiver, and processor of information. More specifically, it evolved to
receive and process information from the environment — the surrounding
Grid — and then to make judicious decisions regarding behaviour based on
the results of its computations.

In the last chapter, we defined perception generally as the receiving and
processing of information from the environment. A single cell perceives
its environment via the gamut of receptors embedded in its surface, the
information flowing from the external world to the networks of signalling
molecules in its interior. Organised at a much higher level than the individual cell, your brain is a perceiver *par excellence*, not only receiving information from the environment, via the sensory apparatus, but also
using this information to generate a model or map of this environment.

The world you live within is this model.

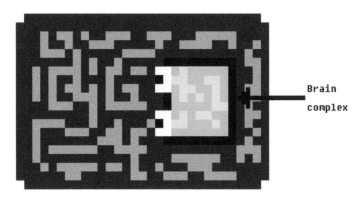

Brain
complex

Environment

**The brain is an information complex specialised in receiving and
processing information from the environment [surrounding Grid]. It uses
this information to build a model of the environment.**

If you look around you now, you will see a world rich in information: all the objects, colours, textures, relative distances, and values associated with all these things — beauty, disgust, ambivalence, etc — are encoded by information generated by your brain. Your world is entirely that information experienced from the inside, from *within*. Your brain generates this information, not because it will afford you a beautiful world to enjoy, but because the phenomenal world thus manifested is useful to you and for your continued survival. A world that contains useful information about the ongoing activity in the surrounding environment provides an organism with an evolutionary advantage over those organisms unable to build such a world. The Grid contains a multitude of information complexes with varying levels of complexity — we recognise these as other organisms and inanimate objects, but let's not lose sight of their true nature as emergent structures built from information. Some of these complexes contain information either beneficial or essential to an organism — what we might recognise as food or sunlight — and some contain information harmful to an organism. Predators, for example, are other organisms that take an active interest in destroying us because they consider us food. However, there are more subtle forms of information: the distances between objects enable us to navigate the Grid to seek certain types of information whilst avoiding others, for example. All of this information is encoded within your brain and forms a model of the external world sculpted over millions of years of evolution.

As a brain evolves, so does the world that it generates[2]. The random changes that occur at the genetic level during reproduction can increase the chances that an organism will successfully survive and reproduce — these are called *adaptive* changes. For example, a slightly longer beak might enable a bird to reach insects embedded more deeply in the trunk of a tree, giving that bird an advantage over birds with shorter beaks. This well-fed bird will be more fertile and more likely to produce a number of strong offspring. Changes can also disadvantage an organism and make it less likely that it will successfully reproduce — these are *non-adaptive* changes. This process of evolutionary change also applies to the phenomenal worlds built by the brain. As a brain evolves towards increasing complexity, it becomes capable of building more and more complex phenomenal worlds containing more and more sophisticated information about the surrounding environment, as well as being able to process and make decisions based on this information in more sophisticated ways.

Organisms with brains that generate poor models of the world — perhaps models that fail to distinguish between predators and prey — will be less likely to survive and reproduce than those with brains that generate adaptive models.

Imagine a primitive brain that is little more than a random information generator, receiving and processing very little information about the environment. Whilst this brain might generate a sort of phenomenal world, it is likely to be a highly unstable and chaotic world that could in no way be described as a model of the environment. In other words, the information being generated by such a brain tells the organism almost nothing about the information being generated within the environment. The amount of information shared by two variables — in this case a *brain* and the *environment* — is called MUTUAL INFORMATION and is a measure of how much we can learn about one variable by knowing something about the other. For example, if you show me a photograph of one of a pair of identical twins, without seeing the other, I already have information about the other twin: the colour of their eyes, the shape of their nose, etc. The twins share mutual information.

The mutual information between this primitive brain and the environment is very low: even if you had access to all the information being generated by the brain, you would learn very little about the information being generated in its environment. As a brain evolves, and its phenomenal world changes, it begins building worlds that contain more and more useful information about the environment. In other words, the mutual information between the brain and the environment increases. Of course, building useful models of the world is not simply a matter of accruing more and more information: a brain must be selective, ignoring irrelevant information and selecting pertinent and useful information. In fact, this filtering of irrelevant information is as important as the selection of useful information. More is not necessarily better.

Psychedelic drugs, including DMT, and certain psychiatric conditions, modify the phenomenal world by modifying the information generated by the brain, creating an alternate model of reality. However, we mustn't make the mistake of assuming that the normal waking world in healthy people is somehow the only true or valid reality: all models of reality are equally real in that they are all built from information.

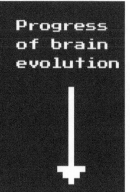

Progress of brain evolution

Information generated by the environment_

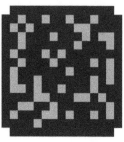

Information generated by the brain_

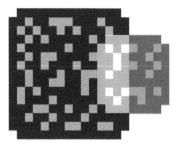

As the brain evolves towards increasing complexity, the mutual information [shown in blue] between the brain's model of the world and the environment increases_

Of course, some models will be more or less adaptive than others, depending on the relationship between the model and the environment. A schizophrenic's world, for example, might well be markedly different to that of the ostensibly sane majority. We can't, however, say that the schizophrenic's world is less 'real' than anyone else's world, only perhaps that it is less adaptive, containing less useful information about the environment and more information unrelated to it — there is less mutual information between the schizophrenic's brain and their environment. The same applies to the altered worlds elicited by psychedelic drugs, which modulate the information generated by the brain and change its relationship to the environment. This change might be subtle, as with a threshold dose of *Psilocybe* mushrooms, or extremely profound, as with a breakthrough dose of DMT.

In summary, your normal waking phenomenal world is a model of the world that your brain has evolved to facilitate your survival as a complex pattern of information on the Grid. Your brain, just like the rest of you, is a highly complex, hierarchically-organised, structure built, like all things, from information, although it sits at the apex of the known hierarchy of complexity. Although all structures generate, process, and are built from information, a brain is special in that it generates information that manifests as your phenomenal world, your personal reality. Whether you are

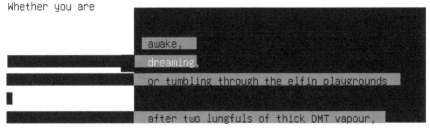

awake,
dreaming,
or tumbling through the elfin playgrounds
after two lungfuls of thick DMT vapour,

your personal reality is always built from information generated by your brain. Although these worlds might be very different in structure and in terms of their relationship to the environment, they are unified by their fundamental nature: information.

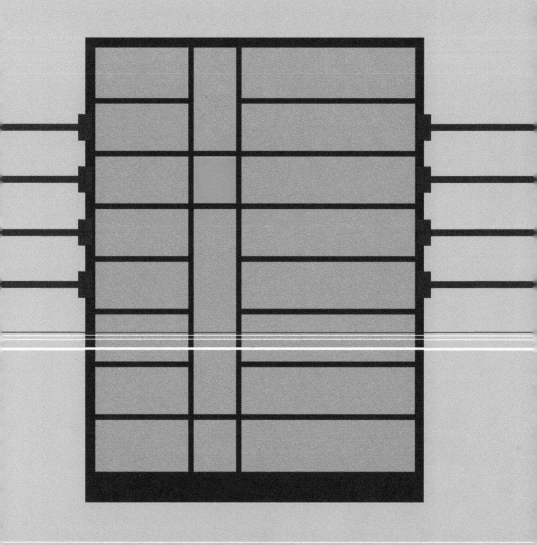

"Far away a crow caws. The earth slowly keeps on turning. But beyond
any of those details of the real, there are dreams. And everyone's
living in them." Haruki Murakami

Chapter 6: How to Build a World
Part I

A brain sits at the apex of the organisational hierarchy with many layers of computational complexity. However, at its most fundamental, like all that manifests in our Universe, a brain is an arrangement of the Cells of the Grid. A brain generates information, as do all things, by the updating of Cell states. These Cells interact locally within the Grid and, through many levels of hierarchical self-organisation, the complex computational properties of a brain emerges. The information generated by your brain manifests from *within* as your phenomenal world. Throughout your life, your brain may generate a number of different types of worlds: the normal waking consensus world of everyday life, the fluid and unpredictable worlds you explore whilst dreaming and, if you ingest certain drugs or plants, worlds altogether stranger than either of those. Whilst clearly distinct, these worlds are unified by their fundamental nature as information generated by your brain.

Although the information generated by a brain is ultimately encoded at the level of the Grid, by the update states of the fundamental Cells, the brain is a hierarchically self-organised structure with emergent information generated at every level. The information that manifests as your phenomenal world is generated at the level of neurons, which are the major information-processing cells in the brain. However, it is important not to lose sight of the fact that

neurons emerge from biomolecules,
which themselves emerge from atoms,
which themselves emerge from subatomic particles,

and so on down to the Cells of the Grid, from which all emerges.

Information is generated within the brain by the activity of its constituent neurons and the interactions between them — electrochemical signals passed between neurons that generate emergent patterns of information. In particular, the neocortex — a folded 2-4mm thick sheet comprising around 50 billion neurons amongst 500 billion supporting cells — is the newest and most functionally-advanced part of the brain and has the primary role in generating the information from which the subjective phenomenal world is constructed. The neurons within the neocortex form an extraordinarily complex pattern of connections and networks that are able to generate collosal amounts of highly complex information experienced as the world.

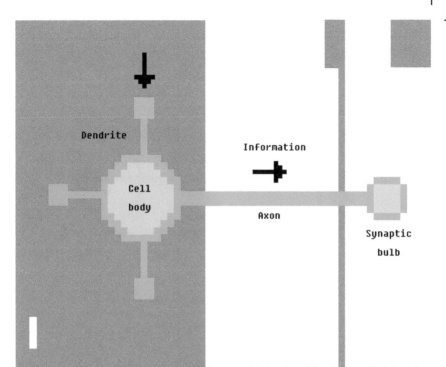

Dendrite

Information

Cell
body

Axon

Synaptic
bulb

Each individual neuron is a highly specialised cell with a distinctive
structure: the central 'hub' of the neutron, the CELL BODY, contains
varying numbers of membrane protrusions — known as DENDRITES and AXONS
— that extend outwards, rather like the limbs of an octopus. The role of
dendrites is to receive information from other neurons and transmit this
information to the cell body, which is where the important computations
take place. Axons, in contrast, have the role of transmitting information
away from the cell body to be passed to other neurons.

Although there are many different types of neurons with particular specialised functions, each has a conceptually simple task: to receive
information, in the form of electrochemical signals from other neurons,
process this information, and then make a decision. The decision is also
simple: remain quiet (do nothing) or fire an electrochemical signal
called an ACTION POTENTIAL. This fire or no-fire decision can be likened
— albeit simplistically — to a two-state system that generates a single
bit of information, with the no-fire decision being equivalent to a '0'
and an action potential being a '1' in digital binary code. When a neuron
fires an action potential, the signal is passed along the axon towards
one or more downstream neurons, which must then decide whether or not to
fire an action potential themselves.

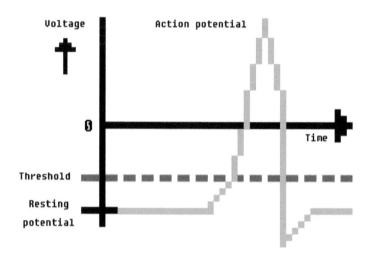

Like all cells in the body, the inside and outside of a neuron is separated by a semi-permeable membrane, which controls the flow of molecules both into and out of the neuron. Neurons are special in that they accumulate certain charged ions — positively-charged *sodium* and *potassium* ions, and negatively-charged *chloride* ions — on the inside and outside of the cell membrane. If these charges are unbalanced, there is a net difference in electrical charge across the membrane, referred to as the MEMBRANE POTENTIAL.

When a neuron is quiet, the electrochemical charge inside the neuron is more negative than the outside — this is known as the RESTING MEMBRANE POTENTIAL. Depending on signals received from other neurons, as well as other neurochemicals, this resting potential can be made more or less negative (lowered or raised). If this potential is raised to a specific level, known as the THRESHOLD POTENTIAL, the membrane potential suddenly and very rapidly reverses — shoots upwards — and then returns to the resting potential. This event is the *action potential* and appears as a *spike* when the electrical activity of the neuron is measured. These spikes are the fundamental 'bit-like' units of information generated by the brain. Sequences of these spikes, like the sequences of 1s and 0s of binary code, are used to encode information.

Neurons do not work in isolation, but are heavily interconnected via their dendrites and axons, forming a bewilderingly complex set of networks among which information — in the form of spikes — is shared, processed, and integrated. Action potentials are initiated close to the cell body, but travel rapidly along the neuron's axon to be passed to one or more dendrites of other neurons. However, the action potential isn't normally passed directly from the axon to a dendrite: There is a small gap between the axon terminal and the dendrite called a SYNAPSE. The role of the synapse is to transmit the information carried by the acton potential from the PRESYNAPTIC neuron to the POSTSYNAPTIC neuron. When the action potential reaches the enlarged axon terminal — the PRESYNAPTIC BULB — a series of biochemical events causes specific molecules — NEUROTRANSMITTERS — to be released into the gap — the SYNAPTIC CLEFT. The neurotransmitter molecules then diffuse across the cleft and bind to special receptor proteins embedded in the postsynaptic neuronal membrane.

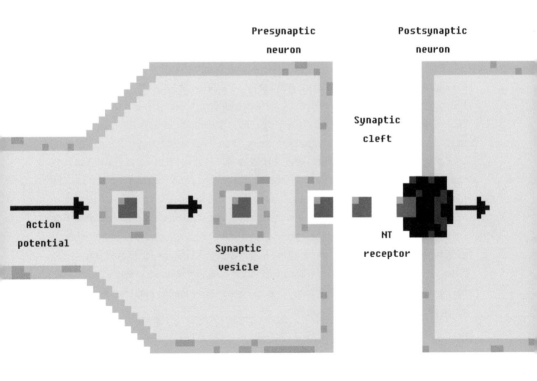

Presynaptic neuron

Postsynaptic neuron

Synaptic cleft

Action potential

Synaptic vesicle

NT receptor

Neurotransmitter [NT]

Dependent on the type of neurotransmitter — there are over 100 different types — and the type of receptor protein, the binding of a neurotransmitter with its specific receptor protein can generate a range of
different effects on the postsynaptic neuron. The most straightforward
effect is to either raise or lower the membrane potential of the postsynaptic membrane. Raising the membrane potential will bring it closer
to the threshold potential, making it more likely that the neuron will
eventually fire an action potential. This is known as an EXCITATORY
POSTSYNAPTIC POTENTIAL (EPSP). Lowering the membrane potential takes it
further from the threshold potential and makes it less likely that a
neuron will fire — this is an INHIBITORY POSTSYNAPTIC POTENTIAL (IPSP).
Since the number of neurotransmitter molecules released from the presynaptic bulb and the number and type of postsynaptic receptor proteins can
be changed, the chemical synapse is highly flexible: the strength of the
synaptic connection can be increased (SYNAPTIC POTENTIATION) or decreased
(SYNAPTIC DEPRESSION) or, in some circumstances, switched off entirely.
This flexibility is crucial for sculpting the information generated by
the networks of neurons in the neocortex and so is of central importance
in building your world.

A single neuron might receive hundreds, or even thousands, of postsynaptic potentials from the axons of other neurons to which it is connected.
Each EPSP nudges the neuron towards the threshold potential, with IPSPs
pulling in the opposite direction away from the threshold. The cell body
integrates all these pushes and pulls and, if the membrane potential at
the cell body reaches the threshold potential, an action potential is
fired. This is how a neuron processes information: it is simply a matter
of whether or not the membrane potential at the cell body reaches the
threshold potential. Whilst each individual neuron can only generate a
single action potential at a time — a single 'bit' of information — the
massively interconnected networks of billions of neurons, each connected
to up to 10,000 other neurons, are capable of generating and processing
colossal amounts of information.

It is this information,
generated by trillions of action potentials per second,
that manifests as your
phenomenal world.

If you look around you now, you will notice that the visual scene you're experiencing has two properties: Firstly, your visual world contains a huge amount of information — observe all the different colours, shapes, and textures in your visual field. Then observe how these take on the form of specific objects that you recognise and how these objects relate to each other in terms of their relative distances from your eyes or how they overlap and interact with each other. All of this information is encoded by the neurons in your brain. In fact, your visual world *is* this information being experienced from your subjective perspective — from *within*.

Secondly, whilst your visual world is extremely rich in information, it is also *unified*. Your phenomenal world cannot be broken down into its constituent parts: whilst the red colour of a coffee mug is clearly distinct from its shape and texture, there is no way to separate them. Your entire visual world appears as a single unified, information-rich, experience. And, every time you move your eyes or a tree outside the window bends in the wind, this unified pattern of information that is your world changes. In fact, every moment of your life, whether you're awake, dreaming, or at the peak of a psychedelic experience, is different from the last. This might seem obvious, but this is only possible because your brain is capable of generating a practically infinite number of unified, information-rich worlds, each different from the last. But how does your brain achieve this?

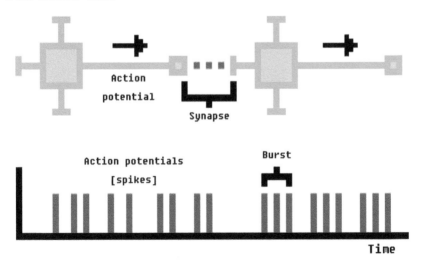

Neurons fire a sequence of action potentials, the rate and pattern of which form a neural code that represents information_

Imagine a small LED bulb that can either be switched ON or OFF. When switched ON, the bulb randomly emits either a RED, GREEN, or BLUE light. Now imagine you possess an extremely simple brain that can only detect the presence or absence of light — more of a light-detector than a brain. Your brain cannot detect the position of the light nor its colour or brightness. With this brain you are only capable of experiencing two different worlds: a world of light or a world of darkness. When the light is switched ON, your neurons fire and your world becomes a world of light and, when switched OFF, you're plunged back into darkness. This is all you can know and experience. The colour of the light makes no difference, since your simple brain is unable to differentiate between colours and can only select between two states.

For you to differentiate between the different colours, we need to build you a more complex brain, by separating the neurons of your brain into three different groups and 'tuning' each group to only fire if a specific frequency of light is detected. When the bulb shines a low-frequency RED light, only the 'RED' neurons will fire, whereas a BLUE light will only stimulate the neurons tuned to that colour. Your brain can now distinguish between the three colours of light and, in fact, you are now able to experience four different worlds: darkness (light switched OFF), plus the three differently coloured worlds of light. Your brain can generate more information — the colour of the light as well as its presence or absence — by being able to select from a larger number of states (refer back to our definition of information from chapter 2). Separating neurons to perform different functions by tuning them to only respond to certain types of information is known as FUNCTIONAL SEGREGATION and is absolutely central to your brain's ability to the generate your phenomenal world.

In the human brain, *functional segregation* refers to the way specific areas of the cortex, and specific sets of neurons within those areas, are responsible for receiving, generating, and processing specific types of information. In the 'light bulb' thought experiment, specific sets of neurons were responsible for detecting specific colours of light. Humans are primarily visual creatures, devoting a large proportion of the cortex to this particular sensory modality. So, it makes sense to explore functional segregation focusing on the visual system. However, these ideas can be extended to include the other types of information — sound, smell, touch, etc — that contribute to your phenomenal world.

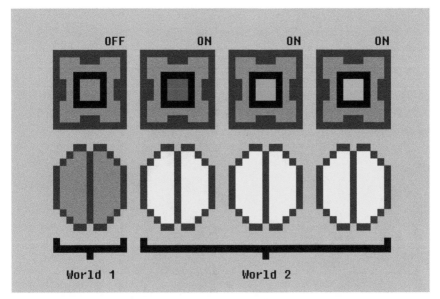

A brain with no functional segregation can only distinguish between light OFF and light ON, but not between the colours of the light_

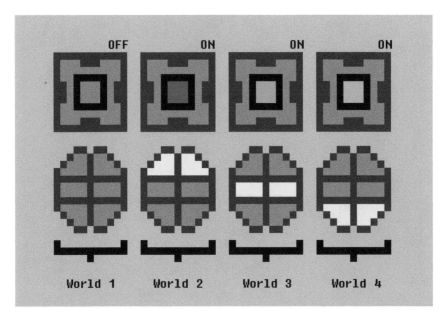

By tuning parts of the brain to only respond to a specific frequency range, the three colours of light can be differentiated_

The areas of the brain responsible for generating visual information lie at the back of the brain in the aptly named VISUAL CORTEX. However, much more extensive areas of the neocortex also make essential informational contributions to your visual world. The PRIMARY VISUAL CORTEX — V1 — sits right at the back of the brain and is the region that first receives visual information from the external world, from the retina at the back of the eye and via a walnut-shaped hub in the centre of the brain called the THALAMUS. V1 is generally responsible for basic visual information, containing 'simple' neurons tuned to respond to certain line orientations or textures, as well as more 'complex' neurons that only respond when a line is moving in a specific direction, for example. From V1, information is sent to the VISUAL ASSOCIATION CORTEX, which contains neurons specialised to represent specific features of the world, such as geometric shapes, colours, or spatial depth. Further downstream, in the TEMPORAL LOBES — which sit at the sides of the brain — are areas specialised for the recognition and representation of certain types of objects, such as faces or animals. Information, in the form of action potentials from the retina and thalamus, is passed along this pathway and spreads to these specialised areas which extract the types of information to which they are tuned to respond. Using this information, the brain constructs your phenomenal world.

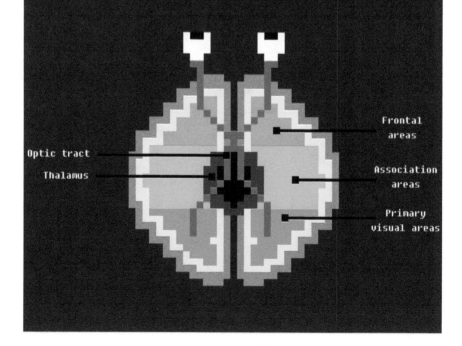

Frontal areas

Optic tract

Thalamus

Association areas

Primary visual areas

To illustrate how this works, let's imagine a highly simplified brain, more complex than the brain we envisaged in the 'light bulb' thought experiment, but still containing only a handful of functionally segregated areas. How would this brain generate a very simple world, containing only a single object: *a smooth red square moving from left to right*? This simple world comprises only a few features that must be encoded in the brain: the form of the square [its edges, corners, and overall shape], its colour, texture, and its movement. Neurons in functionally segregated areas of the brain are specialised to represent these features. An area devoted to processing colour information contains neurons that only respond to specific colours. In this simple world, it is the 'red' neurons that will fire, whilst the 'blue' and 'yellow' neurons, for example, will remain quiet. The same applies to neurons devoted to processing edge, shape, and movement information, with specialised subsets of neurons within each functional area representing a specific feature of the world. Taken together, these neurons form a *pattern of activation* – pattern of information – that represents the moving red square. It's important to remember that, whether or not there is a *red square* in the environment that you are perceiving, or whether the square is a hallucination, dream, or psychedelic vision, its construction is the same. The brain builds the *red square* using information generated by the functionally segregated areas of the cortex. As we discussed in the last chapter, how this red square relates to events in the environment is a different issue entirely, and this might be different depending on whether the square is experienced during normal waking life, during a dream, or during a psychedelic trip.

Of course, the worlds we actually experience are far more complex. However, no matter how complex the world, all of its information must be encoded by a pattern of activation of many different types of specialised neurons spread across functionally segregated areas of the cortex. And, if areas of the cortex responsible for representing specific features of the world are damaged, by a stroke or injury, for example, the sufferer will find himself in a world without those features. For example, the area of the visual cortex called V5 is responsible for the processing of movement information. Damage to this area causes a condition called *akinetopsia* or *motion-blindness*. Individuals with this rare condition live in a world of still images and have no perception of motion. Likewise, damage to the areas responsible for processing colour information results in a monochrome world devoid of all colour.

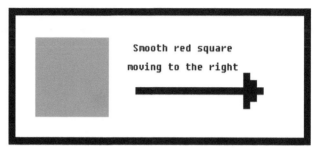

Perceived visual object

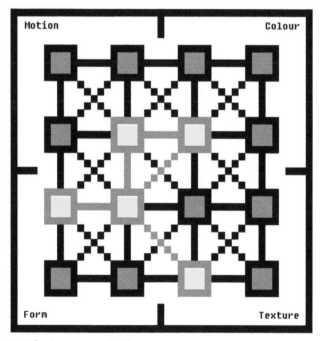

Cortical representation

Inactive neurons

Active neurons

These principles can be extended to areas of the cortex responsible for processing the other types of sensory information and constructing other features of your phenomenal world. For example, a natural sound has a complex wave structure built from simple waves of different frequencies. These waves combine to form the complex wave that stimulates the machinery inside your ear. Specific areas of the auditory cortex are tuned to respond to specific frequencies, and each frequency component of a complex sound wave activates its own frequency-tuned neurons in the auditory cortex. This pattern of activation represents the complex structure of the original sound wave and manifests as the sound you experience.

The smallest functionally-segregated area of the cortex is known as a CORTICAL COLUMN, a cylindrical structure containing about 100 neurons, and the cortex can be described as a mosaic of columns packed side-by-side. At any point in time, the entire cortex displays a complex pattern of activation of these columns[1]. *This pattern of activation is a specific state of the cortex selected from a practically infinite number of possible states and encodes all the information that constitutes your entire phenomenal world at that moment.* In chapter 2 we defined information as being generated when a system selects between a finite number of discrete states. This is exactly how the brain generates information, albeit using a system with a vast number of states. Whenever the cortical column mosaic selects a specific state — pattern of activation — it generates an enormous amount of information by ruling out countless other states.

Patterns of activation of these individual columns encode all the information in your phenomenal world. However, the second fundamental characteristic of your world, after information, is *unification*: your world is always unified. If you look at a bowl of brightly coloured fruits, it's impossible to become confused as to which colour is attached to which fruit. This seems entirely obvious and yet, based on how we understand the brain to function, it's actually quite a feat: The colour information of the fruits is processed at an entirely separate area of the cortex from their shape and, yet, the correct colour is always *bound* to its particular fruit. This is known as the *binding problem*[2], since there is no superordinate area of the cortex — like a projection screen — where all the features of an object are brought together to form the unified structure — the various features of each object remain as a distributed pattern of information across the cortical columns.

Each cortical column in the human neocortex is constructed from six layers: layer I is the outermost layer, closest to the skull, and layer VI is the deepest layer, closest to the centre of the brain. Since these columns are packed together sideways, this gives the entire cortex a six-layered structure.

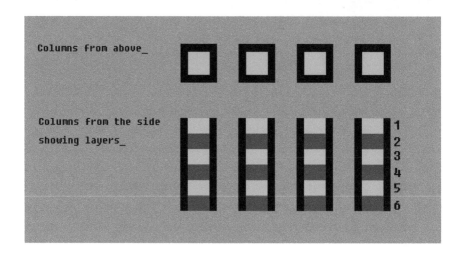

Columns from above_

Columns from the side
showing layers_

1
2
3
4
5
6

The solution to this binding problem lies in the massively interconnect-ed nature of the cortical columns. Rather than a mosaic of independent columns, the columns have dense connections — formed from large numbers of synapses — that allow rapid two-way interactions[3]. So, a pattern of activation of columns can be integrated to form a unified structure. The thalamus, a walnut-sized structure sitting at the centre of the brain, is commonly seen as a relay station through which all sensory informa-tion, barring that from the nose, must pass on its way to the cortex. But this is only part of the story. Each functionally segregated area of the cortex — each cortical column — is reciprocally connected to a corresponding region of the thalamus, forming a THALAMOCORTICAL LOOP. In fact, the thalamus can be described as a miniature map — or 7th layer — of the cortex[4]. So, each cortical column is better described as a THALAMO-CORTICAL COLUMN (T-COLUMN). When a T-column is activated, the electro-chemical activity can be recorded on an electroencephalogram (EEG) as an electrical oscillation at around 40 Hertz, known as a GAMMA OSCILLATION. This particular type of oscillation is important for the integration of T-columns across the cortex[5].

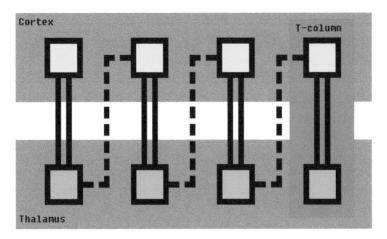

Each cortical column is reciprocally connected to an area of the thalamus forming a loop or T-column. Connections from the thalamus to neighbouring columns allow information to spread between T-columns_

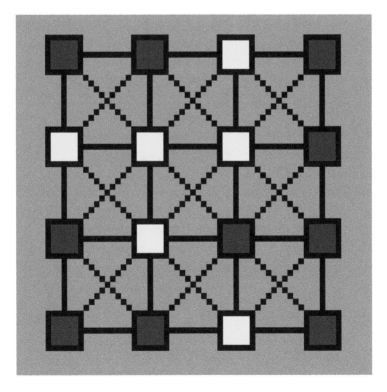

Active T-columns form a pattern of activation that encodes information: a T-state_

When you observe an object, such as a strawberry, for example, the T-columns that encode its various features — its red colour, its shiny mottled texture, its characteristic shape — are activated and form an activation pattern. The electrochemical activity of each activated T-column begins oscillating in the gamma range and these oscillations rapidly become synchronised. This can be compared to the way a wine glass can be made to 'sing', or even shatter, if the right frequency of sound — the natural frequency of the glass — is played. Gamma oscillations can be thought of as the natural frequency of an activated T-column and, when many T-columns are simultaneously activated, they rapidly synchronise their oscillations and transiently self-organise to form a unified structure: a THALAMOCORTICAL STATE (T-STATE)[6]. The activity of a large number of T-columns can be integrated within a few hundred milliseconds to generate the information-rich and unified T-state that encodes your phenomenal world. At every moment of your life, your entire world is a unique pattern of activation of a huge number of T-columns distributed across the cortex and unified by the thalamocortical system. Your world changes from moment to moment as a sequence of these T-states, one T-state flowing into the next. Since the number of possible T-states is vast, when the cortex selects a specific T-state it generates an immense amount of information.

Since all of the information that constitutes your phenomenal world is encoded by these T-states — by the activity of the thalamocortical system — it's natural to wonder what exactly is the role of sensory information from the external world. The construction of a phenomenal world is something the brain does by default and can be fully independent of incoming sensory information. During dreaming, for example, the brain is perfectly capable of building a phenomenal world with all senses apparently intact, despite having almost no access to sensory information. To explain this, we'll first distinguish between two types of information that the brain uses to build your world: the world is itself built from the information generated by the activity of the T-columns that form a unified T-state. We'll call this INTRINSIC INFORMATION. And we'll refer to information that enters the brain from outside, through the senses, as EXTRINSIC INFORMATION[7]. When extrinsic information enters the brain — from the retina or the inner ear, for example — it is not simply added to the intrinsic information. Whether or not there is any incoming sensory information, neurons fire spontaneously and T-columns are activated, generating intrinsic information by forming T-states.

Rather than adding to this intrinsic information, extrinsic information from the senses amplifies or 'awakens' specific patterns of intrinsic information. Specific patterns of extrinsic information are MATCHED to specific patterns of intrinsic information being generated by the thalamocortical system[8]. Or, equivalently, the thalamocortical system can only absorb information that *matches* the intrinsic information it generates. For example, when you look up into a clear blue sky, the blue light activates the *blue cone cells* in the retina, and this extrinsic information is transmitted to the visual cortex as a sequence of action potentials. This pattern of information is matched to the activity of a particular set of neurons in the visual cortex — those tuned to this particular type of information — and amplifies their activity. The effect is to increase the amount of intrinsic information in the T-state that is experienced as the colour blue. Note that the extrinsic information itself never enters the T-state — it can only modulate the intrinsic information being generated by the natural activity of the thalamocortical system.

The thalamocortical system has a repertoire of T-states — patterns of active T-columns — that it tends to adopt[9]. This repertoire is only a subset of the practically infinite number of possible states (we'll see why later). Extrinsic information is *matched* to and so *selects* specific T-states from this repertoire, but never adds to or replaces this ongoing intrinsic information. The brain is not a video camera, capturing moving images of the world and presenting them to consciousness. Sensory information only *modulates the ongoing activity* of the thalamocortical system and your phenomenal world is built entirely from intrinsic information. When you descend into sleep at night, access to almost all extrinsic sensory information is removed and, yet, the brain continues to build complete phenomenal worlds as you dream. These dream worlds usually appear strikingly similar to the waking consensus world. In fact, the only difference between the waking world and the dream world, in terms of their construction, is that the waking world is modulated by extrinsic sensory information, whereas the dream world is not. Without access to sensory information, the thalamocortical system, using T-states from its repertoire, will construct the consensus world as a default. Sensory information *constrains* the construction of your phenomenal world, by selecting specific T-states from the repertoire, but the world is not built from sensory information[10]. Whether you are awake, dreaming, or deep in the DMT worlds, your world is always built from intrinsic information.

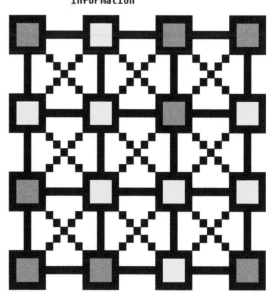

Extrinsic
information

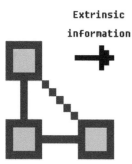

Each blue square
represents a feature
of the sensory data_

Sensory matching

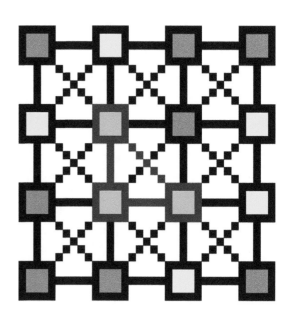

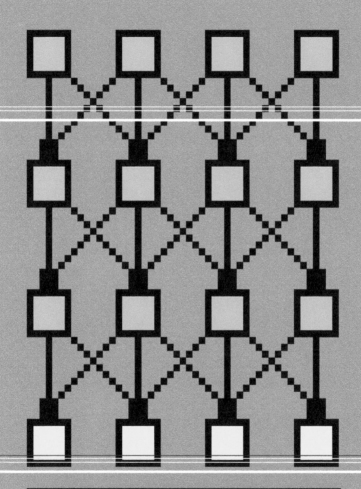

Chapter 7:

How to Build a World Part II

Whenever you are aware of being in a world,

your brain is constructing this world from information generated by its ongoing activity, each moment a pattern of activation of T-columns: a T-state. Your brain moves through these states moment by moment, state by state and, by selecting a single state from a vast repertoire of possible states, generates the massive amount of information that constitutes your entire phenomenal world. Assuming you aren't dreaming, information from the senses reaches the thalamocortical system, is *matched* to this ongoing intrinsic activity and selects T-states from this repertoire, guiding the brain from state to state. Crucially, sensory information only *constrains* the flow of T-states and isn't necessary for your brain to build your world: during dreaming, for example, your brain happily builds your phenomenal world using the same repertoire of T-states it employs during waking. The only difference is that, unguided by sensory information, the flow of states can become rather

erratic and unpredictable,

leading to the often irrational and, sometimes, quite absurd flow of moments experienced during a dream. But dreaming or waking, building a phenomenal model of the world is a skill the human brain has developed over the course of its evolutionary history. Learning to build a stable and richly informative phenomenal reality within which you live out your entire life is the brain's most remarkable achievement, an understanding of which we can glean by thinking again about the purpose of the world built by your brain and the manner of its construction.

The connections between T-columns are essential for generating a unified T-state and phenomenal world. More specifically, connectivity in the brain can be separated into three types[1]:

STRUCTURAL CONNECTIVITY;
EFFECTIVE CONNECTIVITY;
FUNCTIONAL CONNECTIVITY.

Structural connectivity is the physical coupling of neurons using the chemical synapses we met in chapter 6 — this is the brain's basic wiring. Since the neurons communicate using action potentials — spikes — which encode information, it's helpful to think of the brain's connections as controlling the flow of information through it.

These connections form circuits and networks of T-columns that can be stable over long periods of time: from minutes to days or, even, an entire lifetime. However, this wiring is not like that of a computer, since synaptic connections can be strengthened or weakened, removed entirely, or new connections added.

If activity in one T-column causes an effect on the activity in another T-column we say they have *effective connectivity*. Since the thalamocortical system is highly interconnected — *structural connectivity* — information is transmitted between T-columns and, when a T-column is activated, it is likely to activate a number of other T-columns to which it is connected. Of course, effective connectivity depends upon structural connectivity: activity in one T-column cannot affect another T-column unless there is a connection, direct or indirect, between them. Structural connectivity is the wiring scaffold that allows these dynamic interactions and the flow of information between T-columns to occur.

Together with gamma oscillation synchronisation, this effective connectivity manifests as patterns of simultaneous activity of all the T-columns that form a T-state, that forms your world. It is this transient temporal coincidence of activity in separate areas of the cortex — the simultaneously active columns of a T-state — that we refer to as *functional connectivity*.

Functional connectivity is the most fleeting form of connectivity and, whilst depending on structural and effective connectivity, doesn't refer to physical connections between T-columns, but rather to *connections in time*. When, for example, the brain is encoding an object in the visual field, T-columns in the functionally segregated areas that represent the features of that object are simultaneously active: those areas are functionally connected. In a functional MRI image, these areas of the cortex are seen to 'light up' at the same time. All of the T-columns of a T-state are functionally connected since, by definition, those T-columns are active simultaneously. However, unlike structural connectivity, which can persist for a long period of time, functional connectivity can change from moment to moment, as T-states dissolve and are replaced by new T-states. This is the flow of moments, one T-state after the other, that is your experience of living in a world, whether it be this world or an altogether different one.

Structural connectivity_

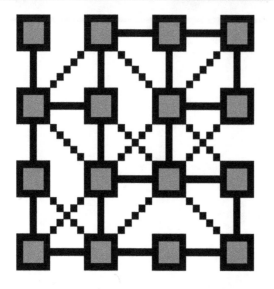

Physical connections between columns_

Changes slowly over days to years_

Shaped by effective and functional connectivity_

Effective connectivity_

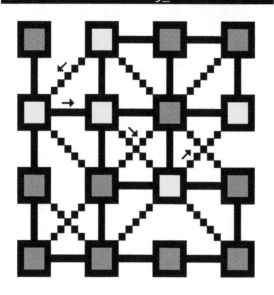

The effect of columns upon each other_

Dependent on structural connectivity_

Effective connectivity causes sets of columns to be active simultaneously: functional connectivity_

Effective connectivity has an essential role in sculpting the thalamo-cortical activity that is your world, as information will tend to spread amongst T-columns that are most strongly connected. Extrinsic information from the external world is always incomplete: when you're gazing out of a window at the world, the brain isn't receiving complete images of the scene that it then somehow presents to consciousness. Your eyes are re-ceiving noisy patterns of light that the retina must convert to action potentials and then pass to the cortex. Once the information reaches the primary visual cortex, it will activate — be matched to — specific T-column populations that are tuned to respond to the features encod-ed in the information received from the retina. These T-columns will then activate T-columns to which they're connected, and the information spreads through the thalamocortical networks dependent on the effective connectivity between the T-columns. After a few hundred milliseconds, the gamma oscillations of the active T-columns will synchronise to form the integrated T-state that encodes your visual experience of the scene. The visual scene is not presented to the brain, but is constructed by it. *Your brain builds your world.*

Sensory information, by activating specific T-column populations, helps to select a particular T-state from the thalamocortical system's state repertoire. But this is only possible because of the connectivity between the T-columns: only a relatively small number of T-columns are activated by the sensory data, but effective connectivity allows this information to spread to other T-columns to form a complete T-state. Which T-state is selected will depend on how these T-columns are connected, as well as on T-columns that were already active when the sensory information reached the cortex from the sensory organs. Only a small amount of extrinsic information is required to select a particular T-state and connectivity does the rest. When you dream, the sequence of T-states that you expe-rience as your dream world is not modulated by extrinsic sensory data. And yet, the dream world normally appears very similar to the normal waking world. The connectivity of the thalamocortical system organises its activity such that the T-states that are adopted tend to be those of the waking world. Your brain knows how to build the consensus waking world and will tend to do so whether or not it has access to extrinsic sensory information. The consensus world is built from the T-states of thalamocortical state repertoire, and the states within this repertoire are determined by the connectivity of the thalamocortical system.

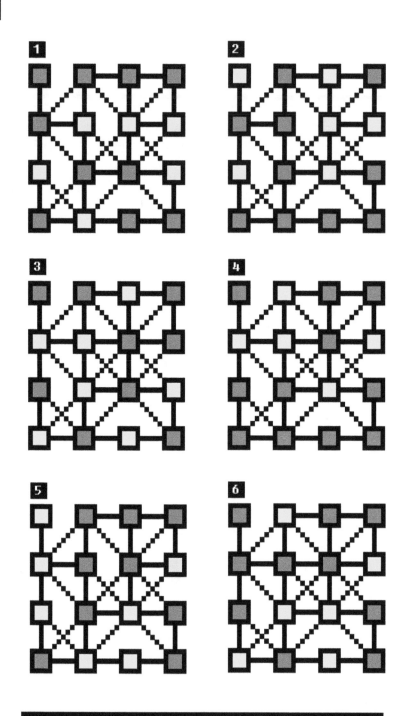

A selection of 6 T-states from the vast repertoire of states

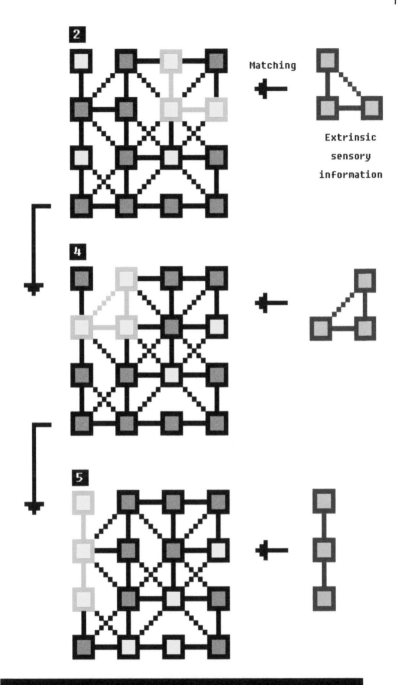

2 Matching

Extrinsic sensory information

4

5

By being matched to thalamocortical activity, extrinsic sensory information selects T-states from the repertoire. During dreaming, the brain moves more freely between T-states without guidance from sensory information_

The human brain was not dropped to Earth as a pristine engineered world-building machine — the brain evolved to build a model of the world. The brain is essentially an information-generator, the information being generated by the intrinsic activity of the thalamocortical system. The sequence of T-states adopted by the brain encodes the informational structure of your phenomenal world. The modern human brain builds the familiar consensus world by default, with T-states being selected from a restricted state repertoire controlled by its connectivity. But how was this repertoire of states formed? How did the brain *learn* to build the world we are all familiar with? On a cellular automaton grid, one can imagine an information complex evolving to receive, process, and store information about activity in the surrounding grid. Your brain *is* this information complex that has evolved on the 3(+)-dimensional Grid of our Universe, and the world it builds serves that same purpose: to generate useful information about the surrounding Grid. The success of any particular model of the world is measured only in terms of its usefulness: does the model make it more likely that an organism, such as yourself, will survive to reproduce? If so, then brains that build such a model will be selected by evolution, whereas brains building poor models of the world will be consigned to the scrap heap of failed attempts.

How does a brain refine its model of the world throughout the course of evolution?

By modifying its connectivity.

The activation patterns — T-states — adopted by the thalamocortical system are determined by its connectivity. This repertoire of states is not fixed, but changes as the connectivity changes. A brain with purely random connectivity might generate as much information as your own brain, although this information would say nothing about the external world — the surrounding Grid. Sensory information would flood into such a brain, cascading through the randomly-connected T-columns and generating an extremely information-rich, but completely meaningless and useless, phenomenal world. In chapter 5, we saw how this corresponds to a lack of mutual information between the brain and the external world [surrounding Grid]. As a brain evolves, the mutual information between the external world and the information generated by the brain increases: *by modifying and tuning its connectivity, the brain builds better worlds that are more informative about the environment.*

The brain's connectivity is shaped over two very different timescales: you are born with a basic connectivity that is shaped genetically, a result of the blend of genes you received from your parents. In addition, your brain's connectivity changes during development – this is also partly controlled by your genes. It is this genetically-encoded connectivity that has been moulded by evolution over countless generations[2]. As the human brain evolved, those brains generating more and more useful and information-rich models of the world were selected for. On a much shorter timescale, your brain's connectivity changes as a result of experience.

From the moment you're ██████████████████████████████████████

██

████████████████████████ dragged ███

████████████████████████████ screaming ███

██████████████████████ from the womb, ███

██

and even before [see chapter 16], your brain is flooded with information through the senses. This information activates T-columns, which then pass the information to other T-columns via the billions of synaptic connections in the brain.

Synaptic connections are special in that the more a connection is used, the stronger it becomes, whereas connections that aren't used may disappear entirely. So, the continuous stream of information entering the brain is not only stimulating the formation of T-states,

██ but is actually sculpting the connectivity of the brain and so ██
██ modifying the repertoire of T-states available to the brain. ██

It is through this repertoire of states that your brain moves, state by state, as it builds your phenomenal world. By selecting from the countless possible activation patterns of the columns of the thalamocortical system, through a combination of evolution, development, and experience, the human brain has learned to build a stable, informative, and useful model of the world. This is the only world your brain knows (or, at least, ought to know) how to build. And such is the proficiency with which your brain builds this world, it performs this task effortlessly, even in the complete absence of sensory information from the external world, as during dreaming.

Weak,
non-specific,
poorly organised
connectivity

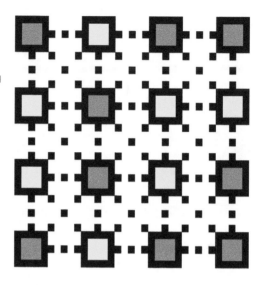

Sensory information ↓

Strong,
highly specific,
organised
connectivity

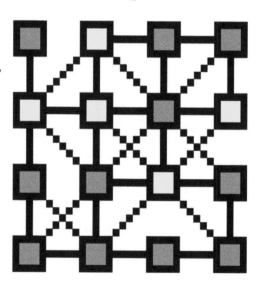

Sensory information is absorbed by the brain and shapes the connectivity of the thalamocortical system. Eventually, the intrinsic activity of the system builds a stable model of the world as a default state_

The worlds that appear when you dream are not mere suggestions or sketches of the waking phenomenal world, but mimic it in every way. The dream state, like the waking state, is characterised by synchronised gamma oscillations and the activation of sensory-specific areas of the cortex[3]. Seeing a face in a dream activates the same areas of the cortex as seeing that face in waking life.

The brain builds worlds in exactly the same way during dreaming as it does when you're awake, and there is no difference in the intrinsic information generated in either state.

The only difference is that, when you're awake, the world is modulated by extrinsic sensory information, whereas this information is excluded during dreaming. Despite this paucity of sensory information, the brain remains active in building the dream world, using the repertoire of T-states developed during evolution, development, and experience. However, since these states are not constrained by extrinsic information, the dream world can become bizarre, often impossible: Faces of family members become simultaneously associated with distant friends or the family dog, whilst the scene shifts inexplicably from the garden at the front of the house to the inside of an aircraft. However strange the dream world might become, it is almost always a more fluid version of the consensus waking world. Of course, the same cannot be said for the worlds into which DMT admits access. The alien worlds that immerse the DMT user bear no relationship whatsoever to consensus reality, with a degree of complexity and strangeness far beyond either the waking world or the dream world. However, since the DMT user enters a phenomenal world, no matter how bizarre it might become, it must be constructed from intrinsic information generated by the thalamocortical system. To understand the nature of these worlds of such ineffable beauty and peculiarity, we must first examine in some detail how psychedelic drugs in general affect the way your brain builds your world.

"A lot can be said for the infinite mercies of God, but the smarts of a good pharmacist, when you get down to it, is worth more."

Philip K. Dick

Chapter 8:

Psychedelic Molecules and the Brain

Neurotransmitters are the chemical messengers charged with transmitting information from neuron to neuron and throughout the brain. Released from the presynaptic bulb, these specialised molecules diffuse across the synaptic cleft to reach the postsynaptic membrane. Upon arrival, by binding to protein receptors embedded in the postsynaptic membrane, they can have a variety of effects on the postsynaptic neuron. Over 100 natural neurotransmitters have been identified, each with their own particular roles in brain function. Most types of neuron will only secrete a single type of neurotransmitter, with glutamate, dopamine, acetylcholine, and serotonin being some of the most important.

Neurotransmitters are stored inside small bubbles of membrane — *synaptic vesicles* — in the presynaptic terminal. When an action potential reaches the presynaptic terminal, a sequence of biochemical events is triggered causing the synaptic vesicles to fuse with the presynaptic membrane, which releases the neurotransmitter into the synaptic cleft. A synapse is usually a tight, one-to-one, connection between a presynaptic bulb and a postsynaptic membrane — neurotransmitters are unable to diffuse out of the synaptic cleft and potentially affect other neurons. This allows synapses to form the precise wiring of the brain — its *structural connectivity* — and, as such, is referred to as WIRING TRANSMISSION.

In contrast, other synapses are much more open — with a wider synaptic cleft — and allow the neurotransmitter to diffuse out of the synapse and have effects on large numbers of neurons at the same time. This is known as VOLUME TRANSMISSION and the neurotransmitters involved are given the name NEUROMODULATORS to distinguish their role from that of the *wiring neurotransmitters*. Each type of neuromodulator can bind to a specific set of receptors, each having a characteristic effect on the neuron in which it's embedded. For example,

serotonin [5-hydroxytryptamine, 5HT]

can bind to at least seven different types of serotonin receptor, each of which has a very particular effect. So, serotonin's effect on a neuron is determined entirely by the types of serotonin receptor it possesses, if any at all. A typical neuron in the brain will contain receptors for a number of different neurotransmitters and neuromodulators, perhaps with several different subtypes of each.

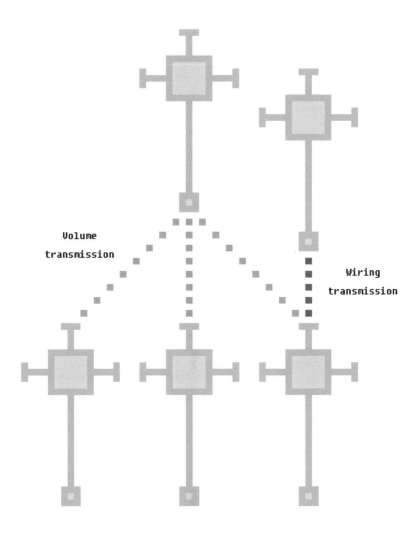

Volume transmission

Wiring transmission

■ Narrow synaptic cleft means neurotransmitter provides a direct connection to the postsynaptic neuron_

■ Wide synaptic cleft allows neurotransmitter to diffuse widely and affect a large number of postsynaptic neurons_

One of the most common effects of a receptor is to alter the membrane potential of the neuron. If the membrane potential is raised closer to the threshold potential — known as DEPOLARISATION — then the neuron is more likely to fire an action potential, since the sum of EPSPs resulting from presynaptic activity is more likely push the membrane potential over the firing threshold. This is also called EXCITATION, often described as making the neuron more *excitable*. Some receptors have the opposite effect, lowering the membrane potential — known as HYPERPOLARISATION — and pushing it further away from the firing threshold. This makes the neuron less likely to fire, or less excitable. Since the effect of a neuromodulator is determined entirely by the receptors to which it binds, the same neuromodulator might have very different effects on different neurons. And, since neurons usually contain many different types of receptor, it can be difficult to predict the overall effect of a neuromodulator on any particular type of neuron.

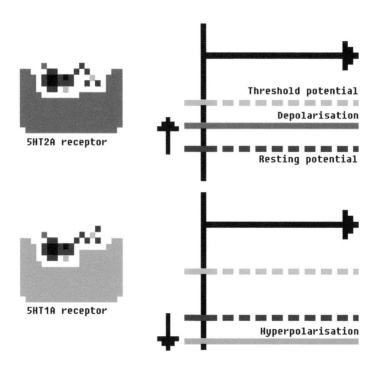

5HT2A receptor

Threshold potential
Depolarisation

Resting potential

5HT1A receptor

Hyperpolarisation

The serotonin [5HT] receptor subtypes, 5HT1A and 5HT2A, have opposing effects on the membrane potential_

Serotonin [5HT] is the most important neuromodulator with regards to the effects of the classic psychedelics, which include LSD, psilocybin, and DMT. Serotonin is secreted exclusively by small clusters of neurons at the base of the brain called the RAPHE NUCLEI. Although the Raphe nuclei form a small structure, the axons from their neurons spread out like long tendrils that can reach almost every area of the cortex. Serotonin has a number of roles in cortical function, but we'll be focusing on one only: its effects at a type of neuron known as a PYRAMIDAL CELL. These neurons form the main cortical component of the thalamocortical loops that are so important in building your phenomenal world. Pyramidal cells send their axons from *layer 5* of the cortex to the thalamus, and receive input from the thalamus in return, completing the loop. They get their name from the triangular, pyramid-like, shape of their cell body, with dendrites pro-truding from the apex of this pyramid and projecting high into the upper layers of the cortex. These APICAL DENDRITES receive the inputs from the thalamus, as well as from other types of neurons surrounding them in the cortex. Serotonin binds to specific receptors embedded in the membrane of these apical dendrites.

There are seven recognised classes of serotonin (5HT) receptors — 5HT1 to 5HT7 — with some of these classes also containing subtypes. For example, the 5HT2 receptor class contains three receptor subtypes: 5HT2a, 5HT2b, and 5HT2c, each with its own particular effects. The most important site of action of the classic psychedelics is the 5HT2a receptor[1], and the potency of a psychedelic drug correlates quite closely with how strongly it binds to this particular receptor subtype[2], and blocking this receptor abolishes any psychedelic effects[3].

Serotonin binding to the 5HT2a receptor has a *depolarising* effect on a pyramidal cell — the membrane potential is nudged towards the threshold potential. Serotonin also binds to 5HT1a receptors — also found on pyram-idal cells — but this has the opposite, *hyperpolarising*, effect on the neuron: activation of this receptor pulls the membrane potential further from the threshold, making it less likely that the neuron will fire. Since these two different receptors share space on the same neuronal mem-brane, they have an antagonising effect on each other, with the 5HT2a re-ceptor exciting the pyramidal cell and the 5HT1a receptor inhibiting it[4]. As such, the balance of 5HT2a vs 5HT1a activation sets the excitability of the pyramidal cell and, by extension, the entire cortex.

The Thalamocortical Loop

An axon from a thalamic neuron projects to the apical dendrites of a
cortical pyramidal cell, which sends its axon down towards the thalamic
neuron, completing the loop_

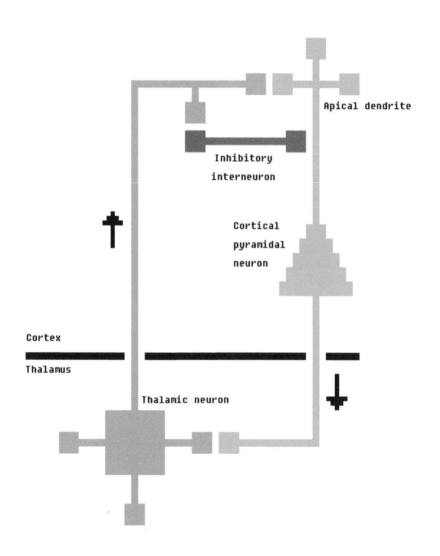

Apical dendrite

Inhibitory
interneuron

Cortical
pyramidal
neuron

Cortex

Thalamus

Thalamic neuron

As well as their antagonising effects on neuron excitability, the 5HT2a and 5HT1a receptors also have opposing effects on gamma oscillations, which are important for integrating the pattern of T-column activation to form a unified T-state. Activation of the 5HT2a receptor *promotes* gamma oscillations, whereas 5HT1a receptors *inhibit* these oscillations[5]. Under normal circumstances, it is serotonin that occupies and activates both receptor subtypes, tuning the excitability of the cortical pyramidal cells and setting the balance of cortical activation. The significance of this balance can be appreciated when it is disrupted.

The classic psychedelics bind selectively to the 5HT2a receptor, but have little activity at the 5HT1a receptor subtype. This tips the balance in favour of depolarisation, exciting the cortex and promoting gamma oscillations in thalamocortical loops. This has two effects: firstly, the cortex becomes more sensitive to incoming sensory information — the spikes that reach the cortex via the thalamus are likely to activate a larger number of T-columns than they would in the absence of the drug. The reason for this is straightforward: whether or not a pyramidal cell, and by extension a T-column, is activated depends entirely on whether the excitatory postsynaptic potentials (EPSPs) it receives push its membrane potential over the firing threshold. When sensory information, in the form of action potentials, reaches the cortex, large areas of the cortex, and so large numbers of T-columns, will receive this information. However, most of these columns will not be activated, since the EPSPs will fail to push the membrane potential of the pyramidal cells over the threshold. By binding selectively to 5HT2a receptors, psychedelic drugs set the *basal membrane potential* — the potential in the absence of stimulation — of all the pyramidal cells slightly higher. This means that more of these cells will be nudged over the threshold as the sensory information reaches the cortex. Furthermore, once these T-columns are activated, they are more likely to successfully transmit their information to other T-columns, since they will also be more excitable.

The overall effect is that sensory information is not only better absorbed by the primary sensory areas of the cortex, but is also more likely to spread to other areas of the cortex. The architecture of thalamocortical connectivity usually controls the spread of information between columns, but this control begins to slip as the T-columns become increasingly excitable and more readily activated by even relatively weak connections.

By selectively binding to 5HT2A receptors, psychedelics depolarise the pyramidal cell, increasing its excitability by pushing its membrane potential closer to the firing threshold_

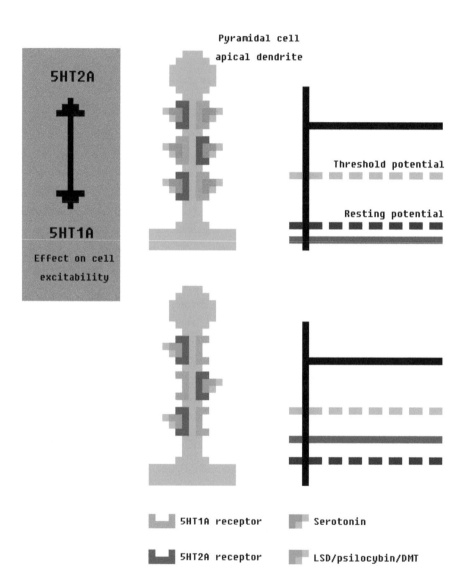

Pyramidal cell apical dendrite

5HT2A

5HT1A

Effect on cell excitability

Threshold potential

Resting potential

5HT1A receptor

5HT2A receptor

Serotonin

LSD/psilocybin/DMT

Secondly, the enhancement of gamma oscillations by psychedelics means that activated T-columns are more likely to be incorporated into an integrated T-state, which contains all the information that constitutes a phenomenal world. Furthermore, since this enhanced gamma effect is widespread across the cortex, highly coherent gamma oscillations are likely to spread more freely, potentially even in the absence of incoming sensory information[6]. T-columns are recruited into novel activation patterns that aren't part of the normal repertoire of T-states[7] — the T-state repertoire expands to include entirely new states.

Normally, the intrinsic activity of the thalamocortical system provides the context for incoming sensory information, which is *matched* to this ongoing activity, selecting and amplifying states from the T-state repertoire. However, in the presence of a psychedelic drug, the inflated state repertoire means that sensory information may select completely novel T-states.

This sequence of novel T-states is experienced as a profound change in both the structure of the phemonenal world and the way the world flows from moment to moment. The world shifts from being stable and predictable to being unstable, unpredictable, and novel. Colours appear brighter and richer, the boundaries between objects appear to blur, as objects blend into one another or reconfigure their structure and identity before your eyes: The hose-pipe on the lawn morphs into a coiled snake or the pebble driveway transforms into a bed of gleaming jewels. There might even be a blending of the normally well-demarcated sensory systems, perhaps with visual areas of the cortex being recruited in response to sound information: blue flashes accompany the dog barking across the street, or music from the stereo system elicits bursts of coloured light that shimmer across the visual field.

When a psychedelic molecule enters the brain, the world appears to change and, indeed, it does change: the activation patterns of the T-columns have changed, and this means the information generated by the thalamocortical system — the information from which your world is built — has changed. Again, we return to the idea that your phenomenal world is built from information. When this information changes, so does your world.

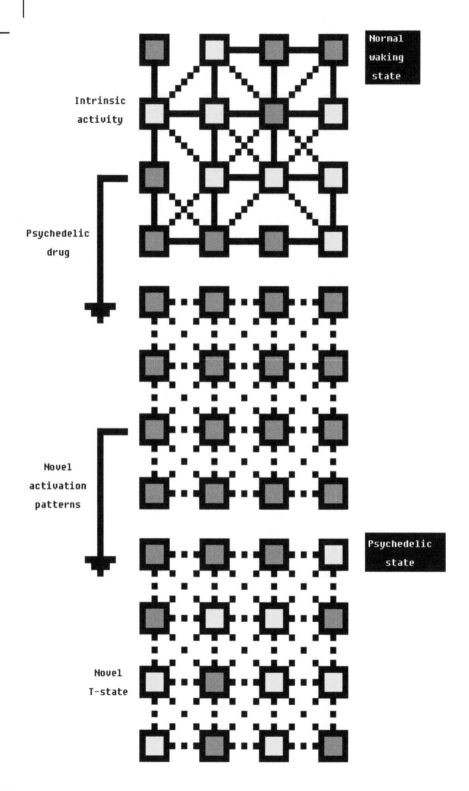

These effects on cortical activity can be visualised using modern brain imaging techniques, such functional magnetic resonance imaging (fMRI), which allow the activity within the brain to be measured and monitored in real time, producing a visual image of activity in the various areas of the cortex. The connectivity of the brain is organised into networks, many of which are common across all healthy people. The brain areas that comprise these networks tend to be activated together and are associated with specific functions. For example, the so-called DEFAULT MODE NETWORK (DMN) comprises several brain areas and their connections, mainly located towards the midline of the brain, that are activated when a person is focused inwardly rather than on the outside world or on any particular task. Hence this network is also known as the *task-negative network*. Daydreaming, ruminating about the past or future, or thinking about one-self tend to be associated with activity in this network. TASK-POSITIVE NETWORKS, on the other hand, are outward-looking networks activated when an individual is actively engaged in a specific activity that requires attention, such as solving a maths problem or driving. Strong connections within these networks help to organise brain activity and restrict the flow of information between cortical areas. Monitoring the activity with-in these well-defined networks provides a measure of how well the brain's information is organised. When an individual is given a psychedelic drug, as predicted by their effects on pyramidal cells, these networks appear to break down: Activity ceases to be kept within the order of the networks and flows more freely between different types of network[8,9]. Overall, the cortical activity appears more disorganised and random, which is the visual expression of the novel T-states generated by psychedelics.

As a brain evolves, it becomes better at generating a useful and inform-ative model of the environment. This can be quantified as an increase in the mutual information between the brain and the external world. Psychedelics appear to temporarily reverse this process: the intrinsic activity — information — becomes looser and spreads more freely under the influence of a psychedelic drug. The information generated by the thalam-ocortical system becomes less constrained by sensory information and the mutual information between the brain and the external world decreases. This doesn't mean that the psychedelic state, or the phenomenal world experienced, is any less valid or 'real' than the normal waking world, simply that the world has changed its structure. The thalomocortical system explores novel T-states that fall outside its normal repertoire.

In the NORMAL WAKING STATE, activation of the DMN and TPN are well-demarcated and anti-correlated. As the DMN is activated, the TPN is strongly suppressed, and vice versa_

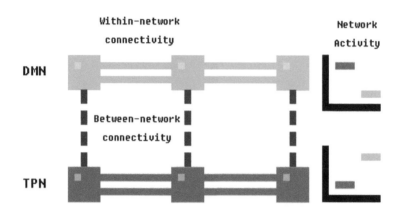

Within-network connectivity

Network Activity

DMN

Between-network connectivity

TPN

In the PSYCHEDELIC STATE, there is a loss of differentiation between the DMN and TPN networks. Information begins to flow between normally separated networks, indicated by an increase in between-network connectivity on functional MRI_

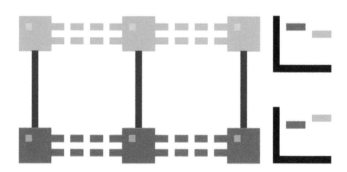

It's possible that the brain actually generates more information during the psychedelic state than during the normal waking state. However, less of this information will be immediately useful from an evolutionary perspective. The brain is as much concerned with ignoring or filtering out information not considered useful in the immediate concerns of survival as it is with selecting important information. Since sensory information is not simply swallowed by the brain but, rather, selects T-states from the thalamocortical system's state repertoire, information that doesn't *match* this ongoing activity has no effect on the brain and is effectively filtered out. However, since psychedelics expand this repertoire to include completely novel T-states, a broader range of sensory information will happen to match this activity, including information that would normally be filtered out. As a result, the brain becomes better at absorbing sensory information and the world becomes far richer as the thalamocortical system progresses through a series of novel T-states. However, this richer, expanded, and more flexible state of consciousness comes at a cost.

The brain must strike a balance between order and disorder [like other complex systems, its dynamics sit at the *edge of chaos*]: the organisation of information using networks is essential for building a stable and predictable world that can be used to make judicious decisions about behaviour. Locating food and avoiding predators, for example, requires the brain to know the difference. However, if the networks militate a too stringent and inflexible form of order, then the potential for creative thinking, incorporating new ways of looking at the world, or simply reacting rapidly to the ongoing influx of sensory information would be compromised. At the opposite extreme, the complete disintegration of network organisation would yield a highly flexible state of consciousness with the potential for immense creativity and novelty. However, such a brain would completely fail to organise the contents of the world into meaningful objects about which astute decisions could be made — the world would be utterly chaotic and confusing. By relaxing the order imposed by cortical connectivity, psychedelics shift the brain towards disorder and generate a richer and more flexible world — without descending into chaos — with the potential for greater levels of creativity and novel thought than the normal, undrugged, state. However, a significant amount of order must be sacrificed and the psychedelic state is perhaps suboptimal from an evolutionary standpoint — at least in the long term.

LSD, psilocybin, mescaline, and DMT are the 'big four' classic psychedelics, each built from either a tryptamine or phenethylamine nucleus,
but each with its own characteristic way of affecting the brain and
changing the phenomenal world. However, the effects of DMT on the structure of the world are far more dramatic than those produced by normal doses of LSD, psilocybin, or mescaline. All psychedelics, including the many
novel drugs derived from the classic psychedelics, modify the information
generated by the brain and, in so doing, modify the world. Usually, the
world that manifests under the influence of a classic psychedelic is an
altered version of the consensus world. DMT, however, is exceptional:
given a sufficient dose — around 30-50 mg for an average person — the
world is not changed but, rather, replaced entirely. Whereas the other
psychedelics partially reduce the mutual information between the brain
and the environment, DMT reduces this information to zero. DMT is a 100%
reality channel switch: the DMT worlds bear no relationship whatsoever
to consensus reality.

As a psychedelic drug reaches the brain, cortical networks begin to break
down and lose their control over the spread of information through the
cortex. Increasing the dose of a psychedelic tends to enhance this effect
even further: a larger number of drug molecules reach the brain and bind
to a larger number of 5HT2a receptors. This increases the excitability
of pyramidal cells even further and, consequently, the intrinsic activity of the thalamocortical system becomes even more fluid, unstable, and
unpredictable. At its extreme this can lead to a complete disintegration of the phenomenal world and utter confusion, with the user tumbling
into a maelstrom of fragmented forms with no point of reference in the
outside world. All sensory information enters the brain freely, without
constraint, activating an apparently boundless variety of novel T-states
without any relationship to each other or the environment. This confusion
is typical of the early stages of a DMT trip. However, once this early
phase passes, the DMT worlds no longer resemble maelstroms of confusion, but stable worlds of crystalline clarity. Whilst the DMT space is
undoubtedly of immense complexity, it possesses a character all of its
own and doesn't result from random neural activity. The DMT worlds are
thoroughly alien, often indescribably bizarre and, yet, possess a striking number of characteristic features commonly reported by large numbers
of users. Before discussing how DMT achieves this, let's look at these
strange worlds in more detail.

Chapter 9:

An Introduction to Hyperspace

"I like to think that I am a rigorous
        thinker and, yet,
        here I am telling you that
        elf legions await in hyperspace,
        one toke away…"

                    Terence McKenna

Outside of laboratory studies, LSD, psilocybin, and mescaline are usually
ingested orally. LSD, owing to its extreme potency, is usually prepared
by soaking absorbent blotter paper, perforated into square 'tabs', in a
solution of the drug, which is then dried. These tabs allow the minute
quantities that constitute a fully active dose to be easily measured out
for consumption. Psilocybin and mescaline are most often ingested in
their natural form: psilocybin by eating any of the many varieties of
mushroom of the *Psilocybe* genus, and mescaline by chewing the dried tops
— 'buttons' — of the *peyote* cactus.

Despite being present in an abundance of plants, DMT cannot be consumed
orally in its natural form. Monoamine oxidase A (MAO-A) is an enzyme
present in the gut that is important for metabolising certain amino acids in food. In particular, MAO-A selectively breaks down molecules that
contain a single amine group (hence, *mono*amine). Of course, this also
includes DMT, which is rapidly destroyed by MAO-A on entering the gastro-
intestinal tract. However, DMT can be rendered orally active by consuming
a drug that temporarily suppresses MAO-A activity — a MAO inhibitor or
MAOI — allowing DMT to enter the bloodstream and reach the brain. This
drug combination is the basis for the traditional Amazonian brew known as
ayahuasca which, in its minimal form, is a decoction of two plants, one
of which contains DMT and the other a MAO inhibitor. The *ayahuasca* brew is
extremely bitter and unpleasant, with most users struggling to gulp down
the nauseating dark brown liquor before enduring several hours of violent
purging. This limits its popularity as a means of ingesting DMT outside
of traditional shamanistic ceremonies and, by far, the most popular mode
of DMT ingestion is via the pipe. Freebase DMT is readily vaporised with
gentle heat, usually in a small glass pipe, and a full dose can be inhaled
in one or two lungfuls. Care must be taken not to burn the drug, which
not only destroys it but also produces a highly noxious vapour often
described as tasting like burning plastic and making inhalation without
coughing extremely challenging.

Amongst DMT aficionados, there is much debate over the most efficient means of vaporising DMT, which is seen as something of an art. A jet flame torch lighter is ideal, since it burns with a hot sootless flame. Regular butane lighters produce large amounts of soot, which coats the pipe and obscures the DMT as it vaporises, making it easy to burn.

The most common practice is to empty the lungs fully, before slowly inhaling between one and three lungfuls of the DMT vapour, with the final lungful being held as long as possible to maximise absorption into the bloodstream. Or, some suggest the best approach is simply to inhale as much as possible, as quickly as possible, until holding the pipe becomes impossible.

A quiet and comfortable environment, usually indoors, is preferred, as is somewhere to lie down. Whilst outdoor DMT trips are not unheard of, a breakthrough trip is almost always experienced from behind closed eyelids with little opportunity for enjoying the natural world, so a comfortable bed in a dimly-lit room is as good a place as any. The onset of the trip is both rapid and overwhelming, usually beginning before the user has exhaled the final lungful, and at which point the eyes are closed and the user lies back and holds tight.

Initially, the voyager is hurtled through a rapidly changing procession of complex visual imagery — described by Timothy Leary as like being "fired out the muzzle of an atomic cannon with neo-byzantine barreling" — and often accompanied by a distinctive metallic buzzing or whirring sound as the drug takes hold. If the dose is sufficient, this complexity eventually gives way and the user bursts through a veil or membrane — sometimes heralded by a teeeeeeeeeeeeeeeeeeeeeeeeeeeeeeeeeeearing or

popping sound

— into a completely novel domain of impossible dimensionality and teeming with entities of immense intelligence and power. Many DMT trips falter before reaching this 'breakthrough' phase, and the user is dragged back into the consensus world without having reached the other side of the veil. However, if the dose is properly prepared, the vaporisation technique properly honed, and the lungs well-seasoned, entry into the DMT hyperspace is assured.

During normal waking life, your phenomenal world is constructed by your brain as a model of your environment: the surrounding Grid. In the same way, *hyperspace* refers to the *phenomenal world* experienced during a DMT trip, and is a model of a higher-dimensional environment to which DMT gates access. In later chapters we will dicuss the structure of this environment, and its relationship to the Grid, in great detail.

Once breakthrough is achieved, transfer to hyperspace is rapid and complete, as if the consensus world has been switched off and an entirely new world switched on. Users typically describe this thoroughly alien world as being more real than ordinary waking reality and the lucidity of the experience is striking, with trippers typically able to experience the bizarre effects as if in an ordinary waking state. Not all DMT users enter the same type of world and, of course, we shouldn't expect that everyone thrust into an alternate reality ought to have exactly the same type of experience: if an alien was dropped onto a random place on Earth, his experience would depend on the geographical location he happened to touch down upon. Landing on a busy street during rush hour in Hanoi would be incomparable to landing on a barren freezing Siberian tundra, and reporting back to his alien kin he would describe two very different worlds. The regions of the DMT hyperspace that users find themselves tumbling into is largely, at least in inexperienced users, beyond their control, as are the types of entities encountered. With experience, however, some users can learn to direct their journey towards specific areas of hyperspace and into meetings with certain types of intelligent entity.

Entry into hyperspace, especially the first time, is almost invariably accompanied by a feeling of overwhelming shock and astonishment. This is a normal reaction — these worlds are not just strange, but ineffably bizarre and seemingly impossible in their complexity and construction. Most users describe an unshakeable feeling of absolute authenticity and the undeniable presence of extreme intelligence beyond anything that could be experienced in the consensus world. The apparent impossibility of these worlds and their contents stems partly from two characteristic features of the DMT hyperspace that distinguish it from consensus reality: inordinate complexity and the perception of higher spatial dimensions (i.e. beyond three). This complexity doesn't manifest as unbridled chaos or random configurations of bright colours and geometric forms but, in the words of author Graham Hancock[1], these worlds are:

highly artificial_

constructed_

inorganic_

technological_

There is an undeniable sense that these realms are not merely novel domains of the mind, but

> alien habitats constructed by a hyperintelligent hand,
> complete with the jewelled cityscapes and whistling machinery
> of a highly advanced alien society.

The ferocity of the initial entry phase into hyperspace will often overwhelm even the most seasoned traveller, and neophytes are advised against trying to make sense of their new hyperdimensional surroundings or to control the experience. At least for the first few journeys, it is advisable to relax as much as possible and simply observe.

Often the construction of the DMT space rapidly transcends the merely remarkable and moves into the unambiguously impossible: the traveller is transported into realms of apparently higher-dimensional structure, or presented with objects that defy the geometrical constraints of our Universe. It isn't unusual for trippers to recount seeing objects from all sides at once, or observing additional spatial dimensions beyond the usual three. The direct perception of higher-dimensional (i.e. above three spatial dimensions) objects is not possible within our 3-dimensional reality. In fact, such objects are difficult to envisage at all, and the experience of doing so is almost always confounding. A 3D reality is subsumed by any higher-dimensional system, in the same way our 3D world subsumes a lower dimensional one, such as a 2D 'flatland' world. This not only provides an important clue as to the nature of the DMT world and its relationship to ours, but also as to the true structure of our brain complex. We will explore this more deeply in the chapters that follow.

In addition to their inordinately complex structure, the hyperdimensional DMT worlds are made all the more compelling by their occupants. Just as a sprawling alien cityscape would reveal the nature of its architects and residents before a single soul was seen, so the presence of supreme intelligence is felt from the earliest stages of the trip. Once they make their appearance, entities range from savage insectoid and reptilian aliens to benevolent amorphous beings of light. But, by far, the most famous denizens of these fantastical realms are the spritely, mischievous beings often described as 'elves'. Terence McKenna's expositions on these highly animated little creatures, which he dubbed 'machine elves', are legendary:

> "Trying to describe them isn't easy.
> On one level I call them
>
> self-transforming machine elves;
> half machine,
> half elf.
>
> They are also like self-dribbling jewelled basketballs,
> about half that volume, and they move very quickly and change.
>
> And they are, somehow, awaiting.
>
> When you burst into this space,
> there's a cheer!"

McKenna also called them 'tykes', which perfectly captures their playfully impish nature. Whilst ubiquitous, they appear in a variety of forms, ranging from amorphous balls of colourful light to the classic elves of Celtic folklore. Despite this variability, they seem to be unified by their playful character. The elves will often vie for the attention of the tripper and delight in demonstrating their skills, such as singing impossible hyperdimensional objects into existence or leaping in and out of the tripper's chest with much glee.

Whilst the the elfin ones are some of the kookiest occupants of the DMT space, for many, they are generally seen as little more than a distraction from the more powerful beings that inhabit these realms. A comprehensive survey of the fauna would include a bewildering array of creatures[2]:

insectoids,
reptilians and serpents,
elves, imps, goblins, and jesters,
aliens, humanoid and otherwise,
robots and cyborgs,
spirits, angels, demons, gods,
and many other beings that defy categorisation.

For any experienced voyager, it is clear that DMT hyperspace is not a luminal realm populated entirely by beneficent gods of light, but an extremely complex, varied and vast hyperdimensional ecology populated by creatures that vary as much in their character and intent as their outward form.

The interactions between the tripper and these beings are, more often than not, positive. Often, the tripper will find himself being carried or guided by a particular entity acting as a wise elder or protective spirit guide eager to import profound insights into the nature of reality — insights most trippers struggle to carry back into the consensus world. Occasionally, the more lively entities appear simply to delight in the opportunity to show the visitor around the wacky circus. Of course, not all interactions are positive and whilst violent aggression is, fortunately, not that common, non-human entities with some degree of malevolent intent or, at least, that are either visually objectionable or performing some unpleasant act on the user are not infrequently encountered. The curiosity of an entity, which might initially be expressed by a gentle probing can sometimes progress to something more invasive. Outright violence or maliciousness is rare but possible, and it takes both experience and a strong constitution to deal with entities manifesting in this way.

One of the most uncanny types of experience, reported by a striking number of DMT users, is not only the sense that the entities were expectant of their visit, but of a great celebratory uproar upon breaking through into the space:

> "They kept saying welcome back and words like:
> the big winner, he has returned, welcome to the
> end and the beginning, you are The One! As I
> looked around the room I felt the sense of some
> huge celebration upon my entry to this place.
> Bells were ringing, lights flashing..."[3]

This is often accompanied by a profound sense of déjà vu, the unshakeable feeling that one has been there before. In chapter 16, we will discuss exactly why this occurs.

Although the incomprehensible strangeness of the DMT hyperspace is astounding, the existence of such worlds is not the most astonishing revelation afforded by DMT. After all, even the most conservative of physicists would struggle to rule out the possible existence of parallel worlds inhabited by advanced alien intelligences. No, the most astonishing revelation is not the existence of such worlds, but that we have the ability to access them with such facility: by inhaling a couple of lungfuls of one of the simplest and most common molecules in the plant kingdom. It seems impossible to fathom how, when perturbed by such a simple molecule, the human mind can reach into such parallel dimensions of reality and meet intelligent beings that exist entirely independent of us. However, we must realise that DMT was embedded in our reality for precisely this reason: to enable us to access the hyperdimensional realm lying orthogonal to ours. Rather than an alien realm, this is the realm from which we have become alienated, and the realm to which we will ultimately return: *resolution of the Game*. To fully understand how this can be achieved, we first need to understand the relationship between our Universe and the DMT hyperspace.

He thought he'd seen treebranches in a yard beyond the window. Filled with small figures waiting for he. Wizened and crouching, barbate and cateyed dwarfs with little codpieces of scarlet puce.

Cormac McCarthy

Chapter 10: Information Flow Through the Grid

Vaporisation of DMT in a small glass pipe remains the simplest, most reliable and, consequently, most popular mode of entry into the DMT space. Within seconds of inhaling a lungful of DMT vapour, almost before the pipe leaves the lips, the DMT molecules flood the brain and the world begins to change. If the dose is well-measured and the vaporisation technique adroit, breakthrough into the hyperdimensional habitat of the machine elves will occur only a few seconds later. Entry into the DMT space has only one absolute requirement:

> the flow of information from normally inaccessible dimensions of reality into the brain.

DMT changes the information generated by the brain such that it can no longer be modulated by — no longer *matches* — information being received via the usual sensory apparatus from our lower-dimensional Universe [Grid], but instead begins to match information being received from the DMT space. The brain *loses* the ability to sample information from our Universe, but *gains* the ability to sample information from the DMT reality. Consequently, the brain stops constructing a model of the consensus world and begins building the DMT world. But the question remains as to how information to which the brain normally has no access can flow from these hidden dimensions into the brain. Before we can deal with this interdimensional information flow, we must first deal with the flow of information within the usual dimensions of our reality.

Reality is built from information, and the complex forms that fill the world are complex patterns of this information, self-organised in an emergent hierarchy of complexity from the level of the Grid to the level of living and conscious organisms. The interactions between 'objects' in the world are the interactions between patterns of information. Everything is information and its processing. When these patterns of information interact, information is processed, information flows. If a *glider* on a 2D cellular automaton scuttles across its grid and collides with another critter, the information from which the *glider* is built — it is, after all, just a pattern of information — flows into the other critter, in a completely literal sense. The direction of flow is somewhat ambiguous, but we'll simply refer to this as SIDEWAYS INFORMATION FLOW: the flow of information within the same organisational level of a complex, hierarchically organised system.

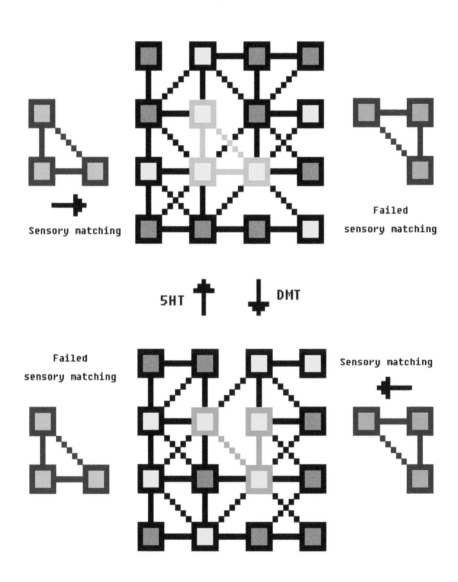

Regular sensory information

Information from alternate DMT reality

Sensory matching

Failed sensory matching

5HT   DMT

Failed sensory matching

Sensory matching

When a photon of light is absorbed by an electron, causing it to leap to a higher energy level within an atom, there is a flow of information from the photon to the electron. In fact, there is nothing other than a flow of information, and the electron's quantum numbers change in response to this: *the electron computes its next state*. In chapter 3, we discussed how molecules inside living cells form complex networks of interactions. When a protein, or other type of molecule, inside a cell interacts with another molecule, it might modify the structure or change the way that molecule operates. This is simply the flow of information between molecules which are themselves — without wanting to labour the point — patterns of information. In fact, the entire workings of a living cell can be described as the flow of information through these complex networks of molecules. Rising to the macroscopic level, we observe the flow of information between neurons and the networks they form in the brain — the activity of neurons, the undulations and spikes of their membrane potential generate information that flows between neurons via the machinery of the chemical synapse. So, although your world is built from information, this information is not static.

Information is dynamic_____

The dynamism of the information from which your world is constructed is never more apparent than during the peak of a DMT trip. Your world is built from information generated by the activity of your thalamocortical system, from the patterns of activation of the myriad columns from which the cortex is built — the sequence of T-states. Psychedelics change your world by changing the activity of the cortical system and so change the information that constitutes your world. *Your model of the external world is altered*. Modern neuropharmacological techniques have revealed the binding of classic psychedelics, including DMT, to the 5HT2a receptor as being primarily responsible for these effects. Activation of this receptor causes pyramidal cells to depolarise, promotes gamma oscillations, and changes the patterns of information generated by cortical activity. These changes in information manifest as a change in the world you experience, which is the psychedelic state.

 Glider  Spaceship

The information encoded
by the glider flows into
the spaceship...

...both are destroyed_

If we examine this process more deeply, we can view it as a flow of information: information flows from the drug molecule to the receptor and, from the receptor, it flows into the network of molecules inside the neuron. This ultimately has the effect of depolarising the neuron, bringing it closer to its firing threshold. But the information flow doesn't stop there: whilst the drug-receptor interaction takes place at the molecular level, the psychedelic effect occurs at the highest level of organisation of the brain: the level of global cortical activity. This isn't particularly surprising or mysterious, but it illustrates an effect known as UPWARDS INFORMATION FLOW[1,2], which occurs when behaviours or processes that occur at a lower level in an organisational hierarchy have an effect at a higher level. This type of information flow can be observed in all self-organised complex systems. In fact, *upwards information flow* is essential for complex systems to self-organise. For example, in a flock of starlings, we've seen how the adherence to a few simple rules by the individual birds — the *bird level* — causes the dynamic behaviour of the flock — the *flock level* — to emerge. In a cellular automaton, such as the Game of Life, we've seen how simple update rules at the *cell level* cause complex, high-order, structures to emerge at levels above the base grid. When a psychedelic molecule enters the brain, the information flows from the molecular level, upwards through the organisational hierarchy, to the level of the integrated T-states that constitute your phenomenal world.

Emergence is precisely the effect of information as it flows through layered complex systems, and describes the layered complexification of the information from which all such systems are built. The relatively simple interactions between large numbers of neurons, the exchange of information between them, engenders the highly complex, adaptive behaviours of the complete brain. We cannot necessarily predict the effect of information as it flows through the increasingly sophisticated levels of a complex system, which will depend on the system it's entering. At every level of the organisational hierarchy information is generated. The form of the flock of starlings contains new information not present at the level of the individual birds. The birds contain the potential to form a flock, but the 'flock information' isn't generated until the flock itself forms. A drop of water contains the potential for every possible snowflake form, but only the potential. The information in the form of the snowflake is only realised when the snowflake manifests, as molecular level information flows to the level of the crystalline form.

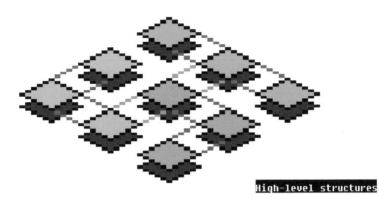

High-level structures

Upwards
information
flow

Downwards
information
flow

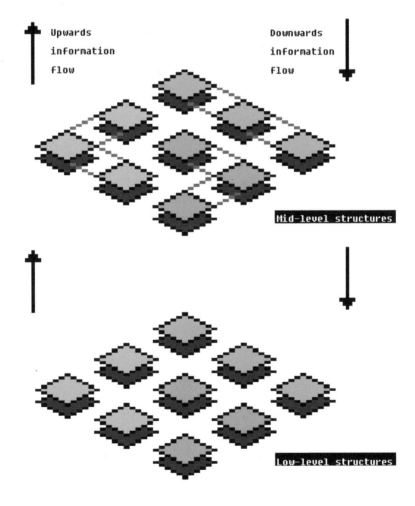

Mid-level structures

Low-level structures

Remember, structures that emerge on the Grid are not static objects, but processes — the dynamic processing of information, updating with every click of time. A complex structure doesn't emerge and then statically maintain itself —

all forms are constantly emerging.

Even an apparently static structure is built from Cells that update themselves with each time step, just as a whirlpool doesn't stop whirling once it forms, save it should disappear. The whirlpool is a dynamic pattern of information, a process, just as the critters that emerge on an automaton grid are dynamic patterns of information. So it's important not to get confused between the idea of information flowing through a complex hierarchical system that has already emerged and the information flow that causes the system to emerge in the first place. *They are the same.*

Information flowing upwards through a complex hierarchy isn't simply passing through —
it is actually generating the hierarchy.

As water flowing downriver gets caught up in a whirlpool, the flow of water that maintains the whirlpool is not distinct from the water that caused it to form in the first place. The whirlpool is a process that is not simply emergent but is constantly *emerging* with each moment, and will continue to do so as long as it manifests. Likewise, *your brain is continuously emerging, together with the world that manifests from behind your eyes.*

psychedelic drugs don't so much change the information generated by the brain as cause a different brain to emerge together with a different world_

Without upwards information flow, self-organisation would be impossible. The information being generated at the level of the Grid flows upwards and causes *fundamental particles* to emerge. The interactions between these particles generate information that flows upwards, causing *atoms* to emerge, and so on upwards through the hierarchy to complex living organisms — information being generated at one level causes the increasingly complex structures to emerge at higher and higher levels.

This upwards information flow approach to understanding complex life is standard: physics underlies chemistry and chemistry underlies biology. However, whilst being standard, it is incomplete. There is an upper limit to the complexity that can be achieved with upwards information flow alone. The new information generated at each level of an organisational hierarchy is not restricted to flowing only upwards to generate even higher levels of organisation. This emergent information — the form and dynamics of the flock or the behaviour of a complex ant society — can also flow in the opposite direction, and this DOWNWARDS INFORMATION FLOW is essential if we are to fully explain the emergence of life and, eventually, the unique properties of DMT.

Downwards information flow occurs when information generated at a high level in an organisational hierarchy has effects on structures or behaviour at a lower level in the hierarchy[3]. Returning to the flock of starlings, the emergence of the flock can be partly explained in terms of upwards information flow:

> the individual birds observe their closest neighbours and adjust their flight speed and heading according to a few simple rules. This causes the characteristic flocking behaviour to emerge — information flows from the *bird level* to the *flock level*.

However, once the flock begins to emerge, the individual birds begin receiving, and responding to, information about the overall structure of the flock: its shape, direction of movement, and their own position within it. This is *downwards information flow*, since the *flock information* is only generated at the *flock level*, but affects the behaviour of the individual birds, helping to reinforce the structure of the flock and augment its elaborately dynamic behaviour.

Downwards information flow exerts its effect by *constraining* the behaviour of the lower-level components of the system. The high-level emergent structure/behaviour constrains the behaviour of the lower-level components from which that structure is built: the form of the flock constrains the flight patterns of the individual birds from which the flock is constructed.

To make it clear why this is described as a downwards flow of *information*, it's helpful to return to our definition of information as being generated when

a system selects between a finite number of states.

All of the possible arrangements — states — of the lower-level components of an organisational hierarchy form a *space of possibilities* or a kind of *state space*. By constraining — or *selecting from* — this set of possible states, high-level information generates information about the lower-level components. Or, equivalently, *information flows downwards through the levels of the organisational hierarchy.*

Imagine a snowflake crystallising out of

a drop

a drop

a drop

of

water.

It is said that no two snowflakes are alike, such is the apparently infinite variety of forms. That single drop of water, the countless water molecules and their innumerable interactions, embodies the potential form of all possible snowflakes. So how is one snowflake form selected from all the other possibilities? As the geometric form of a snowflake crystal begins to emerge — high-level information — it begins to dictate the deposition of further water molecules into the crystal. The growing crystal selects from the space of all possible arrangements of water molecules: information flows from the level of the crystal form to the level of the individual water molecules. In the example of flocking, the form and dynamics of a flock of starlings — high-level information — constrains the flight dynamics of the individual birds — lower level information.

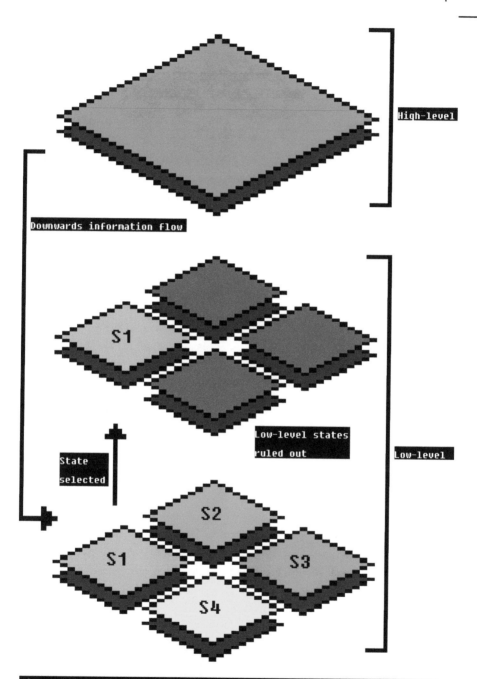

High-level

Downwards information flow

S1

Low-level states ruled out

Low-level

State selected

S2

S1   S3

S4

High-level information selects from the possible states of the lower-level structures, ruling out other states and generating information__ [Note that S1-S4 are not lower-level components of the system, but possible states/arrangements of the entire set of components]

131

In living systems, the role of downwards information flow is elevated from merely modulatory to absolutely critical. Life could not exist without the flow of information both up and down the organisational levels that characterise living organisms. To be considered living, a system must maintain, regenerate, and reproduce itself over time. These are high-level behaviours that can only be performed by a complete cell, dependent on its entire emergent network of molecular components. But, their effects often occur at the level of the individual components themselves: a damaged protein is replaced, a piece of the membrane regenerated, a section of DNA repaired. Information generated at the *cell level* flows downwards to the *molecular level*.

We've already discussed at length how networks of neurons are responsible for generating information and controlling its flow throughout the brain. Certain networks [high-level structures] have algorithmic properties similar to that of computers, allowing them to control specific behaviours. The adoption of a very particular network state — an emergent high-level behaviour that we might experience as a *decision* — might trigger the activation of specific synapses [lower level] that lead to a particular movement, such as turning the head or blinking[4]. This is similar to how a computer software algorithm [high level] controls the switching of transistors [low level] inside the computer to achieve some outcome. In both cases, information generated at a high level of organisation modifies the information being generated at a lower level — information flows downwards.

This two-way flow of information can set up potentially powerful feedback loops: information generated by the low-level components of a system flows upwards and causes high-order information — structures or behaviours — to emerge. This emergent information then flows downwards, constraining the behaviour of the very same low-level components that generated the high-level information in the first place. In certain well-tuned — or evolved — systems, this further enhances or reinforces the high-level emergent behaviour. These *positive feedback loops* can occur at many levels of an organisational hierarchy, helping to stabilise and maintain the entire system and its behaviour over extended periods of time. A living organism is a highly complex system of finely-tuned feedback loops of this kind, from which emerge the dynamic but stable, responsive, and regenerative qualities that characterise life.

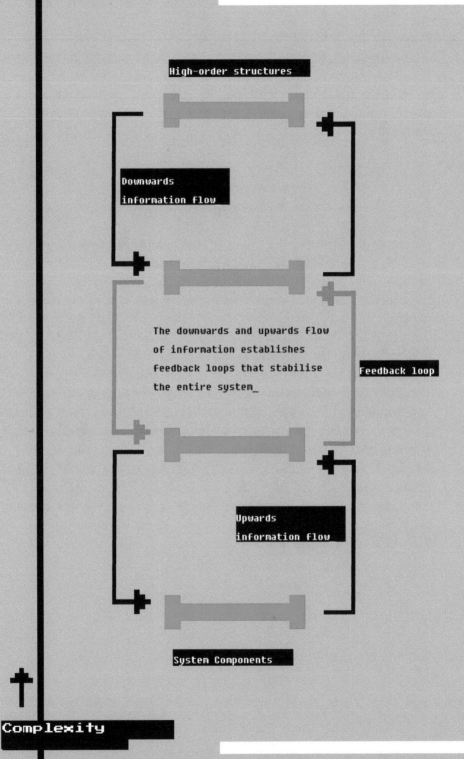

High-order structures

Downwards
information flow

The downwards and upwards flow
of information establishes
feedback loops that stabilise
the entire system_

Feedback loop

Upwards
information flow

System Components

Complexity

Of course, downwards information flow will occur to some degree between any two levels of an organisational hierarchy and, as such, information flows from the very top of the hierarchy to the lowest level, which would be the Grid itself. Your brain and the phenomenal world it generates emerge from the moment-by-moment updates of Cell states at the level of the Grid, with

> new emergent information being generated
> at every organisational level
> from the Grid upwards,

flowing up through the many layers of the complexity hierarchy to the level of global brain activity and consciousness. And, of course, information also flows in the opposite direction, from the level of the coordinated behaviour of large areas of the cortex, down to the level of the Grid.

When the information generated at the cortical level — the information from which your phenomenal world is built — is modulated by the action of a psychedelic drug molecule, this also changes the flow of that information downwards through the layers of organisation to the level of the Grid. DMT, in particular, elicits a highly specific and characteristic effect on this downwards information flow. The result is the gating of information from orthogonal dimensions of reality and the granting of an audience with the elfin folk that dance along the fractal hallways. To understand how this is achieved, we need to think a little more deeply about the relationship between our Universe Grid and the dimensions within which these beings reside.

Chapter 11:

Information Flow Through the HyperGrid

Communicating with the vast array of alien intelligences inhabiting dimensions orthogonal to ours requires, minimally, the flow of information from those dimensions into your brain complex. Whilst everything that manifests in this reality is constructed from information, the brain is special in its role as an exquisite information generator. Indeed, the flexibility and complexity of the information generated by the human brain is unparalleled and, when perturbed by DMT, provides the key to gating information from these normally inaccessible dimensions, experienced as *breaking through* into DMT hyperspace. In the last chapter, we explored how the flow of information both within and between organisational layers is of critical importance in the emergence of complex forms and, ultimately, of living and conscious beings such as yourself. Now we will consider how information flows, not only within an organisational hierarchy, but between dimensions.

So far, we have considered the Universe as a type of cellular automaton: the Grid. A 3-dimensional cubic Grid is the easiest to visualise, since it corresponds most closely to how we view the world. However, we mustn't become too attached to the idea that there is a grid of cubic cells at the ground of reality, since this is inaccurate — in chapter 14 we'll discuss the true structure of the Grid in much more detail. Our Universe Grid is actually part of a higher-dimensional structure which we will call the HYPERGRID, and it is the orthogonal dimensions of this HyperGrid to which DMT gates access. Both the Grid and the HyperGrid [the former being a part of the latter] are generated by the Code. This means that the author of the Code — the Other — exists in a place outside of the HyperGrid.

Imagine a 3D cubic cellular automaton and then take a single 2D *slice* of cells, which itself forms a 2D automaton. In fact, the entire 3D automaton can be considered a stack of 2D slices, and each slice has the same kind of relationship to the 3D automaton as our Grid has to the HyperGrid, in that it is a lower-dimensional slice of a higher-dimensional structure. Likewise, every cell of a 2D slice is connected to the 3D system and, in the same way, every Cell of the Grid is connected to HyperGrid.

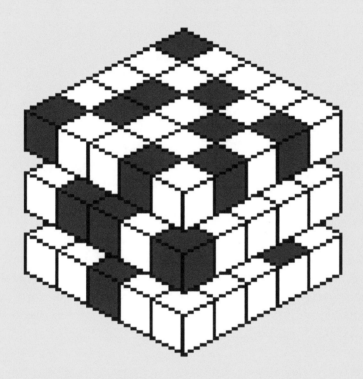

Each 2D slice of a 3D cubic cellular automaton is itself an automaton that may or may not interact with the other slices

However, being structurally embedded in the higher-dimensional system doesn't necessarily mean that information can flow freely from the orthogonal dimensions of the HyperGrid into the Grid. In fact, under most circumstances, our dimensional slice of the HyperGrid remains independent of it. The Grid is a structure that was generated, from the Code, with the express purpose of allowing conscious living beings to eventually emerge, independent of the HyperGrid at large.

> The Grid is a slice of the HyperGrid utilised as a computational device to 'cultivate' intelligences in a lower-dimensional environment informationally detached from the HyperGrid.

However, this detachment is not irreversible and mechanisms exist to gate the flow of information from the HyperGrid to the Grid. Since the entire Grid is a slice of the HyperGrid, your brain is a lower-dimensional slice of a higher-dimensional processor. DMT is an embedded information complex that modifies the information generated by the brain such that information is gated from the HyperGrid into the Grid, allowing the brain to temporarily become a part of this higher-dimensional system. Again, the detailed mechanisms and reasons for this will be explored at some length in later chapters. First, we need to consider how the flow of information between the dimensions of the HyperGrid can be controlled, such that a lower-dimensional slice can be reversibly isolated from the HyperGrid to be utilised as a tool for cultivating emergent conscious intelligences.

The traditional cellular automaton remains a useful way of thinking about the Universe Grid, and it's instructive to assume that the Grid possesses the basic characteristics of such automata:

1. An array of CELLS, each connected to a specific number of NEIGHBOURHOOD cells;

2. Each cell may occupy one of a finite number of DISCRETE STATES;

3. With each click of time, a CENTRAL CELL updates its state based on the states of the neighbourhood cells according to a TRANSITION RULE SET.

The *neighbourhood* of a *central cell* usually comprises the cells in the immediate vicinity, although, in principle, any set of cells can be defined in the transition rules. In a regular 2D cellular automaton, we have already met both the *Von Neumann neighbourhood* and the larger *Moore neighbourhood*, which is used in the Game of Life. So, a neighbourhood is defined as those cells considered in the transition rules. We can also define a neighbourhood as *those cells from which information flows into the central cell*. This is because, in a fully deterministic cellular automaton, knowing the state of a particular cell in a neighbourhood reduces your uncertainty about the next state of the central cell by ruling out specific update states. So, since uncertainty is the opposite of information, the information about the next state is increased – information flows from the neighbourhood cell to the central cell. Of course, if you know the state of every cell in the neighbourhood, then you have all the information required to determine the next state of the central cell with complete certainty, since every update state is ruled out barring the state dictated by the transition rule.

If none of the neighbourhood cell states are known, there is no way of knowing whether the central cell will update to black or white_

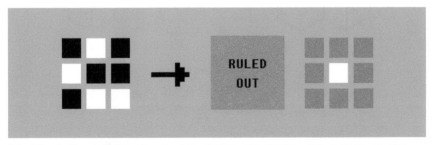

If all of the neighbourhood cell states are known, the update of the central cell is known with absolute certainty [a *bit* is generated]_

It is conceptually simple to extend a 2D cellular automaton, such as the Game of Life, into the third dimension to generate a 3D grid that appears more closely aligned with the world we experience in normal waking life. Of course, in a 3D cellular automaton, the neighbourhood may take into account cells in the third dimension, in addition to the cells in the 2D plane. The 3D version of the Von Neumann neighbourhood adds the two cells directly above and below the central cell to the four cells in the 2D plane, allowing information to potentially flow from the third dimension into the central cell. Extending an automaton into four or more dimensions is also possible, although it becomes difficult to visualise such automata in any straightforward manner. Whereas the most natural way to visualise a 3D automaton is as an array of cubic cells, each cell of a 4D automaton is most naturally represented as a tesseract (or 4-cube): the 4D equivalent of a regular 3D cube (or 3-cube). And, whereas a 3-cube possesses six 2D square faces, the tesseract boasts eight cubic (3-cube) faces. So, in a regular 4D cellular automaton, the 4D Von Neumann neighbourhood would consist of the eight tesseracts contacting each of these faces.

2D and 3D Von Neumann neighbourhoods

Cells in the 2D plane

3D cells orthogonal to the 2D plane

2D

3D

It's not particularly instructive to try and think beyond 4D, although there is no limit to the number of dimensions that a cellular automaton can possess. The important point is that each dimension is *orthogonal* to the others, meaning one can move in any of the individual dimensions independent of the others. Specifying the position of any point in 3D space, or any cell in a 3D cellular automaton, requires precisely three independent numbers, one for each dimension. In our everyday 3D world, orthogonality is synonymous with the *right-angle*, the perpendicular line: a 1D line can be converted into a 2D plane by extending a line perpendicular to it. This plane can itself be converted into a 3D space by extending another line at a right-angle, orthogonal to, the other two. A fourth spatial dimension would be orthogonal — perpendicular — to the other three, and it would be possible to move along the fourth dimension whilst remaining in the same position in the first three. And, defining your location in a 4D spatial world, or the location of a cell in a 4D automaton, would require a four-number coordinate. Also, just as one can take a 2D planar slice of a 3D automaton, one can also take a 3D slice of a 4D automaton, or 3D or 4D slice of a 5D automaton, for example. Our Universe Grid is a lower dimensional slice of a much higher-dimensional structure: the HyperGrid. For simplicity we'll assume the Grid has only three spatial dimensions, but it's certainly possible it contains additional dimensions that we are unable to detect.

For a critter emergent on a 2D cellular automaton, a third dimension orthogonal to the two dimensions of its world would be unimaginable. But, for us 3D beings, it seems perfectly natural, whereas a 4D grid is more difficult to envisage. This difficulty, however, is simply a consequence of our perspective as 3-dimensional beings living on a 3D Grid. Such is the difficulty of visualising automata above three dimensions, purely for explanatory purposes, it will be helpful to reduce the dimensionality of the Grid from three to two dimensions. This will allow us to discuss the relationship between the Grid and HyperGrid without having to visualise 4D space or resort to abstractions. Everything we discuss in this regard applies equally well to a 3D slice of a 4(+)D structure as it does to a 2D slice of a 3D structure. Using this reduced 2D model of the Grid, we will explore how a lower-dimensional slice of a cellular automaton can be reversibly isolated from the higher-dimensional system, and how the patterns of information generated by the slice can be used to gate information from the normally inaccessible orthogonal dimensions.

The purpose of these simple examples is not to provide the actual mechanism of information transfer from the HyperGrid to the Grid. Of course, the actual Grid is not instantiated as such a trivial cellular automaton. However, these examples will provide an intuitive grasp of the way particular patterns of information generated within a lower-dimensional slice of an automaton can be used to control the flow of information between dimensions. We will then apply these general principles to explain how DMT gates access to orthogonal dimensions of the HyperGrid.

If we imagine the Grid as a 2D automaton, not dissimilar to the structure of the Game of Life, then it makes sense to imagine the HyperGrid as a 3D cellular automaton, of which the Grid is a 2D slice. However, since this 2D Grid is an embedded part of the 3D HyperGrid structure, it's unclear whether a conscious intelligence that emerges on this 2D slice will experience a 2D or a 3D world. This will depend on whether or not information flows from the 3D HyperGrid to the 2D Grid, which will be determined by the neighbourhood and transition rules, since the flow of information is controlled entirely by these rules. With a single isolated 2D automaton, the choice of neighbourhood is obviously restricted to cells within the 2D plane, since there are no other cells to consider. And, if the explicit purpose of the Grid was an automaton irreversibly isolated from the HyperGrid, then the slice can be constructed such that the transition rules only take into account cells within the 2D slice. So, cells in the orthogonal — third — dimension have no effect on cell update states, meaning information does not flow from the third dimension, but only within the slice. This approach is no different from simply constructing a 2D cellular automaton, and the fact that it's a slice of a 3D automaton has no effect on its behaviour or on the environment experienced by any critters that emerge within it.

If, however, the aim is not complete isolation of the Grid from the HyperGrid, but to have some control over the flow of information from the orthogonal third dimension into the 2D slice, then a number of techniques can be employed in the construction of the Grid. To understand some of these, we'll first explore how cell states can be encoded that can receive information from a specified number of dimensions. We'll then discuss how the patterns of activity generated by the Grid can be used to trigger the emergence of Cell states that allow the Grid to receive information from normally inaccessible orthogonal dimensions of the HyperGrid.

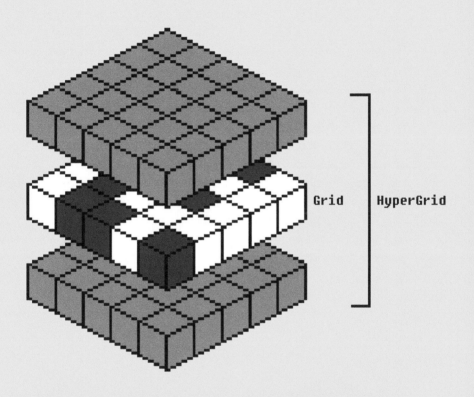

Grid HyperGrid

If information doesn't flow between the Grid and the HyperGrid,
the Grid remains structurally connected to, but informationally
isolated from, the HyperGrid_

To keep things as simple as possible, we'll initially restrict the Grid to two states: BLACK and WHITE, and use the Von Neumann neighbourhood. Firstly, we'll consider the BLACK state only, and think about how we can isolate cells in the BLACK state from the third dimension, allowing cells in this state to only receive information from the 2D Grid. To construct a well-defined standard 2D cellular automaton, for each cell state a rule must be defined for every possible neighbourhood configuration. In the example below, for a cell in the BLACK state, the transition rule set must define how the cell updates for each Von Neumann neighbourhood configuration. Since there are 64 possible neighbourhood configurations, we need to define 64 rules for the BLACK state [the same applies to the WHITE state]. Extending our automaton into the third dimension means taking into account the two additional orthogonal cells to generate the *3D Von Neumann neighbourhood*. Since there are four possible combinations of these cells — BLACK-BLACK, WHITE-WHITE, BLACK-WHITE, or WHITE-BLACK — the number of possible neighbourhood configurations expands by a factor of four: *for every 2D Von Neumann neighbourhood configuration, there are four 3D Von Neumann neighbourhood configurations, meaning four rules must now be defined* — it's helpful to think of each 2D Von Neumann neighbourhood rule as being associated with this group of four rules in the 3D Von Neumann neighbourhood.

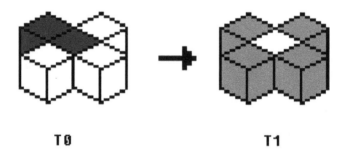

T0   **T1**

This rule dictates that a BLACK cell updates to a WHITE cell with the 2D Von Neumann neighbourhood in this particular configuration [the neighbourhood cells are greyed-out after the update since each of their states will depend on their neighbourhood]:

However, if we extend the transition rule to take into account the orthogonal cells — 3D Von Neumann neighbourhood — we now have four rules to specify. According to these rules, if precisely one of the orthogonal cells is BLACK, then the central cell remains BLACK when it updates. Otherwise, it becomes WHITE:

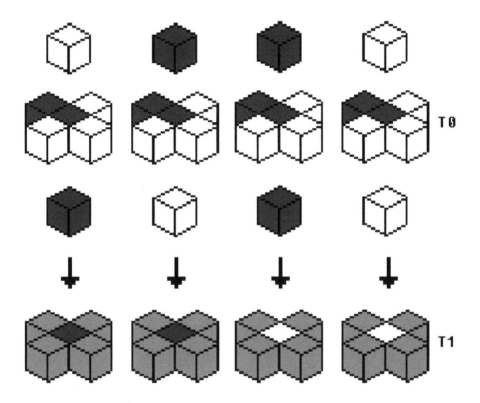

Since the states of the orthogonal cells affect the update of a BLACK cell, information flows from the third dimension into the Grid. This means that, as long as there are BLACK cells present, the Grid is receiving information from all three dimensions of the HyperGrid. If we want to prevent the flow of information from the third dimension of the HyperGrid into Grid cells in the BLACK state, we need to ensure that, *for every 2D Von Neumann neighbourhood configuration, all four of the associated 3D Von Neumann neighbourhood rules generate the same update of the central cell.*

In this second example below, we see the same four 3D Von Neumann neighbourhood configurations. However, in this case, all four rules generate the same update: BLACK to WHITE. This means that, in contrast to the first example, the states of the orthogonal cells have no effect on the update of the central cell. Knowing the states of the orthogonal cells doesn't reduce the uncertainty about the next state of the central cell, meaning *no information flows from the orthogonal cells to the central cell.* We can say that, for this particular 2D Von Neumann neighbourhood configuration, the central cell is INSENSITIVE to the orthogonal — third dimension — cells.

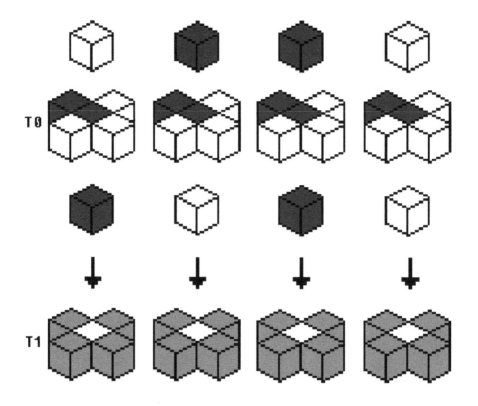

Each of the 3D Von Neumann rules generates the same central cell update, effectively collapsing the four 3D rules into a single 2D rule, since the states of the orthogonal cells have no effect_

Of course, this only applies to this particular neighbourhood configuration: if we want to completely isolate all cells in the BLACK state from the HyperGrid — to block the flow of information from the third dimension — we need to ensure that each of the 64 possible 2D Von Neumann neighbourhood configurations is *insensitive* to the orthogonal cells. That is, we need to construct the rule set such that, for each 2D Von Neumann neighbourhood configuration, each of the four associated 3D Von Neumann neighbourhood rules generates the same update. We can then say that the BLACK state is *insensitive to the third dimension*. If we want to completely isolate the 2D Grid from the HyperGrid, we must construct the transition rules such that the WHITE state, in addition to the BLACK state, is also insensitive to the third dimension. This means that, no matter the state of a cell, it can never receive information from the orthogonal third dimension, and the 2D Grid will always remain isolated from the HyperGrid. As far as any emergent critters on the Grid are concerned, the 2D space of the Grid is all that exists.

We can create a slightly more complex Grid by increasing the number of states from two to three: BLACK, WHITE, and BLUE. And, we'll assume that both BLACK and WHITE states are *insensitive* to the third dimension of the HyperGrid. We will call cell states that can only receive information from two of the three dimensions of the HyperGrid 2-i STATES (i for input). In contrast, let's assume that the BLUE state is *not* insensitive to the orthogonal dimension, meaning BLUE cells can receive information from all three dimensions: 3-i STATES. In the example on the following page, a BLUE cell will become WHITE if precisely one of the orthogonal cells is BLACK, but will remain BLUE otherwise. Since the update of BLUE cells depends on the configuration of the orthogonal cells, they allow the 2D Grid to receive information from the orthogonal third dimension. Of course, in this overtly simplistic example, a 3-i BLUE cell immediately loses its ability to receive information from the third dimension by transitioning to a 2-i WHITE cell. However, this is simply a consequence of using such a trivial, three-state, automaton. More sophisticated automaton structures will enable greater flexibility in the encoding of states to be sensitivie to any particular number of dimensions. The actual Grid, in contrast to our highly simplified model, is a 3D slice of a much higher-dimensional HyperGrid. However, in an analogous manner, Cell states can be encoded in the Grid that receive information beyond the three usual spatial dimensions — we'll refer to these as 4(+)-i STATES.

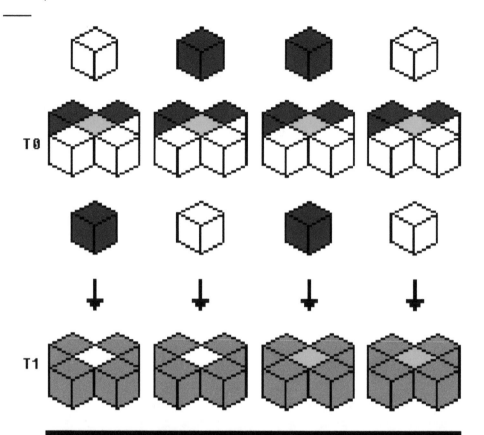

T0

T1

The update of the BLUE central cell depends upon the states of the orthogonal cells, meaning the BLUE cell state is sensitive to — receives information from — the orthogonal third dimension_

The ultimate aim of controlling the flow of information between dimensions of the HyperGrid is to allow certain complex structures that emerge within the Grid — such as conscious brain complexes — to interact with the orthogonal dimensions of the HyperGrid and, eventually, to become part of that higher-dimensional reality. This requires the Grid to be constructed such that 4(+)-i Cell states only emerge within these specific complex structures and under specific conditions — this means structures outside of such structures remain isolated from the HyperGrid. Whether or not 4(+)-i cell states emerge within a brain complex obviously depends upon whether such states are encoded within the rule set. Assuming that states *sensitive* to the orthogonal dimensions of the HyperGrid are indeed encoded, whether they manifest will depend upon whether the appropriate neighbourhood configurations are adopted at the level of the Grid.

In the last chapter, we discussed how information can flow downwards through an organisational hierarchy, affecting the behaviour of structures at a much lower level by selecting from a space of possibilities. Of course, this also applies to a brain complex, the coordinated, emergent activity of which generates information that flows from the level of global cortical activity to the level of the Grid. So, it's possible for information generated at the highest level of organisation of the brain to flow downwards and affect the state configurations adopted by the Cells of the Grid. There is nothing esoteric about this process, which still requires the appropriate neighbourhood configurations to be adopted by the Grid for any particular Cell states to be adopted, but it's prudent to remind ourselves that changes in cortical activity manifested by psychedelic drugs, including DMT, can affect the adoption of Cell states at the lowest level of organisation — the Grid.

So far, we have only considered automata with simple, static rule sets. Whilst the Code at the ground of reality doesn't instantiate such a trivial type of automaton, restricting our discussion this way allowed us to explore the essential properties of cellular automata, and relate these to the construction of reality from digital information, without having to concern ourselves with more complex types of automata. However, to understand how control over interdimensional information flow from the HyperGrid to the Grid can be achieved, we need to consider more advanced techniques, including DYNAMIC RULE SETS and what are known as PATTERN-RULE MAPPINGS. There are many possible classes of transition rules beyond the basic static rules that are specified from the outset and remain unchanged as the automaton runs. In traditional cellular automata, only the *states* of the cells evolve over time, although it is also possible for the *rule set* itself to evolve as the automaton runs.

To give a clear example, it makes sense to return to the simplest type of automaton: the 1D *elementary cellular automaton*, in which each cell updates based on its own state and those of its two immediate neighbours. As explained in chapter 3, there are precisely 256 possible rule sets, each given its own number — from 0 to 255 — according to Wolfram's coding system. The standard way to run an elementary automaton is to select a rule set, say 110, and allow the automaton to run using this set. However, it's also possible for the rule set to change as the program runs, depending on some measurable property of the automaton at each time step[1].

For example, we could set up an ECA with 24 cells and begin with a random configuration of states. At each time step, the number of BLACK cells is counted and the rule set corresponding to that number is applied to generate the update of the cell states. So, if there are 13 BLACK cells, for example, then Rule 13 is applied to update the cell states. Then, the number of BLACK cells in this updated state is counted and the appropriate rule applied, and so on:

This is known as a SELF-REFERENTIAL UPDATE RULE[1], in that a property of the entire system — the number of BLACK cells — determines which rule, selected from a rule space, will be applied to itself. So, instead of a single cellular automaton with a single rule set, we have created an automaton that selects from a space containing 25 rule sets as it runs. We could also design an automaton whereby certain state patterns, rather than simply the number of cells of a certain state, trigger the adoption of specific rule sets according to some PATTERN-RULE MAPPING. Again, this is easiest to illustrate using a 1D automaton.

According to Wolfram's coding system, each of the 256 possible rule sets for an elementary cellular automaton is given a number from 0 to 255. Equivalently, this number can be represented in binary notation as an 8-bit string of 1s and 0s. Rule 175, for example, is

10101111

in binary notation, which is much easier to handle computationally.

Let's create a 24-cell elementary automaton with a random initial state. We can then segregate the 24 cells into eight 3-cell blocks. Ultimately, we will use a property of each of these 3-cell blocks to generate a single binary digit of the new rule. There are eight possible 3-cell patterns, each of which we will map to either a 0 or a 1. There are a number of ways to do this, but we will choose to count the number of state transitions within a block. For example, BLUE-BLUE-BLACK has a single transition (from BLUE to BLACK), whereas BLACK-BLUE-BLACK has two transitions (BLACK to BLUE to BLACK). If there is precisely one transition, then we'll map the block to 1. Otherwise, it maps to 0.

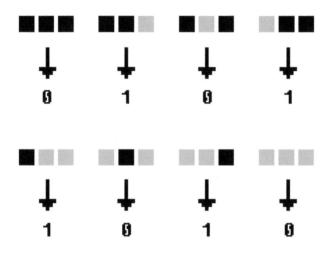

So, overall, the 24-cell pattern maps to an 8-bit binary string: one of the 256 ECA rules. We then apply this rule to the automaton and repeat:

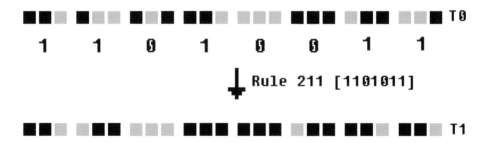

So, at every time step, the pattern generated by the automaton is used to generate the rule used in the next time step. As the cell states of the automaton update, so does the rule set used to perform the update. Of course, using a 3-cell pattern to generate the rule is arbitrary and, particularly in larger and higher-dimensional automata, a vast, but finite, choice of possible patterns and ways of mapping these patterns to the rule space will present itself.

In this example, we used *all* of the cell states — grouped into 3-cell patterns — of the 24-cell automaton to generate the 8-bit update rule, which was then used by all 24 cells. However, it is also possible for patterns generated by a specific number of cells — as opposed to *all* of the cells — to be selectively mapped to a rule set. This generates what is known as a NON-UNIFORM CELLULAR AUTOMATON[2], since the rule set is not uniform across the grid: cells *within* the pattern use the rule set they are mapped to, whereas the cells *outside* the pattern use the original rule set. There is no limit to the number of pattern-rule mappings that can be encoded, although care must be taken to consider how clashes between patterns — when a cell is part of more than one pattern at a single time point — will be handled.

So, in summary, using pattern-rule mappings, the rule set(s) governing the update of a cellular automaton can vary over both *time* and *space* dependent on the patterns of activity generated by the automaton. If a mapping generates cell states sensitive to more dimensions than states reachable using the original rule set, then the mapping provides a mechanism for gating the flow of information from those higher dimensions into localised regions of the automaton.

Applying this to the Grid-HyperGrid relationship, patterns of information generated by the Grid can be mapped to rule sets that specifically generate Cell states *sensitive* to orthogonal dimensions of the HyperGrid, allowing structures expressing such activity — such as brain complexes — to receive this otherwise inaccessible higher-dimensional information.

These *pattern-rule mappings* needn't necessarily be encoded when the automaton is constructed and initiated, but can be programmed into the Grid once intelligent organisms with brain complexes — *candidates for the Game* — begin to emerge.

Specific Cell state patterns
[Perturbation Brain Patterns,
PBP] can be mapped to a rule
set different to that of the
surrounding Grid_

PBP

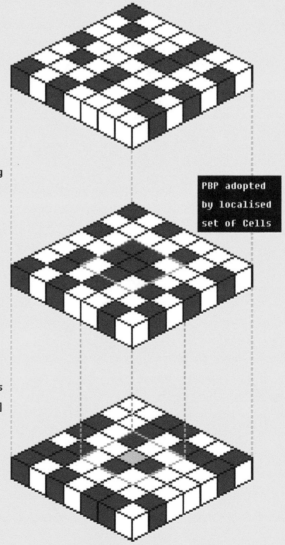

T0

All cells update using
the same rule set

PBP adopted
by localised
set of Cells

T1

Cells within the BPP
update with different
rule set to the
surrounding Grid cells
[pattern-rule mapping]

T2

Since this technique relies on patterns of activity within a brain complex over large numbers of Cells, it's possible to identify Cell configurations characteristic of, and unique to, brain complexes and map these
patterns to rules that will only then be executed within a brain complex.

So, overall, this methodology can be summarised as follows:

The Grid is encoded and allowed to run informationally isolated from the
HyperGrid. This is achieved using a rule set that only instantiates Cell
states that are *insensitive* to the orthogonal dimensions of the Hyper-
Grid. As the Grid runs, the complexificaton of information generated by
the Grid occurs in the manner discussed in earlier chapters. Once brain
complexes are identified as potential candidates for the Game — we'll
discuss later how the Game works — the patterns of Cell activity generated within these brain complexes are analysed so as to construct the pattern-rule mappings that will generate the 4(+)-i cell states within the
brain complex: these Cell states will allow the brain complex to receive
information from the orthogonal dimensions of the HyperGrid.

However, in constructing these *pattern-rule mappings*, NATIVE BRAIN PAT-
TERNS — patterns of brain activity [Cell states] that occur during normal
daily life — are not used directly, since these would cause information
to flow from the HyperGrid into the brain complex constantly, which is
not the desired outcome. Rather, PERTURBATION BRAIN PATTERNS — patterns
of activity resulting from a temporary perturbation of brain activity —
are utilised, such that 4(+)-i Cell states are only generated under this
perturbation. We have already discussed at great length how psychedelic
drugs perturb the information generated by the brain complex, modifying
the way it receives information from the environment and altering the
structure of the phenomenal world. DMT is the molecule embedded in our
reality responsible for perturbing the activity of the brain complex in a
very particular manner, eliciting modified neural information that flows
downwards to the level of the Grid and selects unique patterns of Cell
states. These are the *brain perturbation patterns* that have been mapped
to a rule set engendering 4(+)-i Cell states, thus gating the flow of
information from the normally inaccessible dimensions of the HyperGrid
into the brain.

This is the key to entry into hyperspace.

154

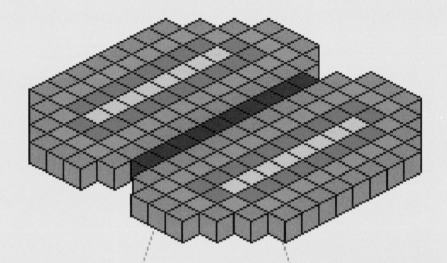

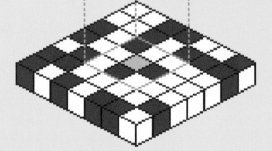

Brain activity perturbed by DMT
generates Perturbation Brain Patterns

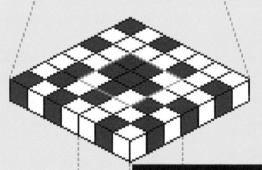

Perturbation Brain Pattern mapped to
rule set generating 4(+)-i states

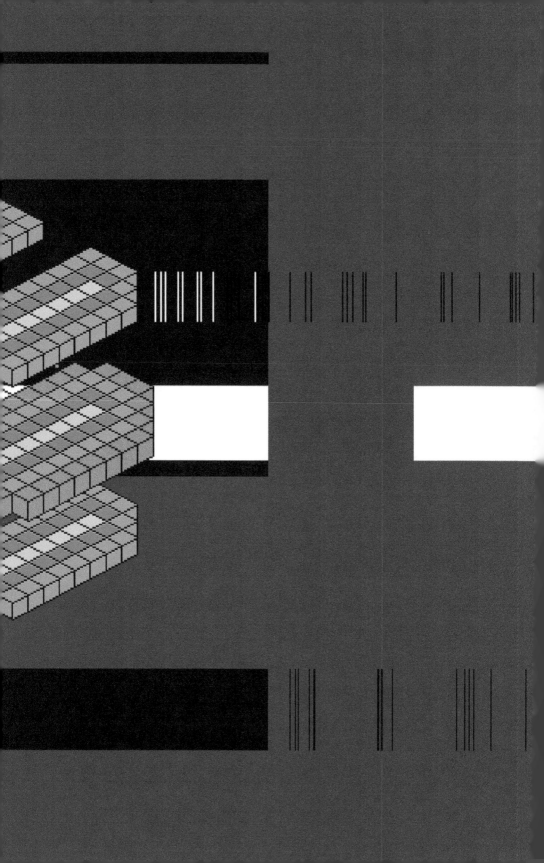

"The Universe is a puzzle. Life is a problem to be solved, it's a conundrum. It's not what it appears to be. There are doors, there are locks and keys, there are levels. And, if you get it right, somehow, it will give way to something extremely unexpected."

Terence McKenna

Chapter 12:
DMT and the Hyperdimensional
Brain Complex

All forms emergent on the Grid of our Universe are unified by their fundamental nature, as *information complexes*: structures built from information itself, instantiated by the Cell states from which the Grid is constructed. From the simplest, first-order, structures to the most supremely complex living organisms with many layers of hierarchical organisation, all are fundamentally Cell-based structures that receive and process information from the surrounding Grid and the other structures co-emergent within it.

This is also true of the HyperGrid, the higher-dimensional system — also generated from the Code — of which the Grid is a lower dimensional slice. It ought to be obvious that the dimensionality of an information complex will correspond to the dimensionality of the slice upon which it emerges and is embedded. Entering the DMT space means confronting a world that is not only startlingly complex, bizarre, and teeming with apparently 'hyperintelligent' living beings, but which is entirely impossible. Indeed, these worlds *are* impossible from the perspective of a conscious being that emerged in a lower-dimensional reality. The information complexes you will meet and interact with in the DMT space — the elves and their curious objects, the impossibly constructed hallways and jewelled temples — emerged, or were constructed by emergent intelligences, embedded in the HyperGrid and, as such, are higher dimensional structures that are beyond inconceivable for a Grid-based organism. Whilst many mathematicians routinely work in high-dimensional spaces, and passable renderings of 4-dimensional forms, such as the 'tesseract' hypercube, give us an idea of the structure of a 4-dimensional object, it's certainly not possible for a 3D brain complex to envisage 5, 6, or 7-dimensional objects and organisms other than in complete abstraction. So, when a machine elf gleefully tosses you a 7-dimensional 'Fabergé egg' across a 9-dimensional room, it's thoroughly confounding.

Whether you are immersed in the mundanities of your regular life in this universe or dancing with Brobdingnagian reptilians in the DMT worlds, the phenomenal world you experience is always constructed by that most complex of all information complexes emergent on the Grid: your brain. A brain is a special type of information complex in that it is adapted to receive, process, and store large amounts of information from the environment: the surrounding Grid. This information manifests as the phenomenal world experienced by the brain.

In earlier chapters, we discussed how your brain generates the intrinsic information from which it constructs your phenomenal world, as well as the manner in which psychedelic drugs can alter this information and change this world. Since this information is largely generated at the level of neurons and their networks, it made sense to set the discussion at this level. However, it's important that we equate the 'wet brain' familiar from neuroscience with the brain as a high-level information complex that emerges on the Grid of our reality. Like all such complexes, the human brain is a structure normally restricted to the Grid's limited dimensionality. The same applies to the phenomenal world it constructs, which we each experience with a dimensionality commensurate with the slice within which we're embedded. This raises the question as to how, in the presence of DMT, the brain suddenly becomes capable of constructing worlds of seemingly impossible dimensionality.

By modulating the patterns of information it generates, and activating the encoded *pattern-rule mappings*, DMT gates the flow of information from the normally inaccessible orthogonal dimensions of the HyperGrid into the brain. As we'll discuss in more detail in the next chapter, this causes the brain to cease building the consensus world and to begin constructing the thoroughly alien, higher-dimensional, DMT worlds experienced as *hyperspace*. To achieve this, the brain doesn't simply passively absorb the hyperdimensional information, but actually undergoes a transformation in its most fundamental, lowest level, structure.

In the last chapter, we saw how the adoption of 3-i (BLUE) states by cells in a 2D slice of a 3D cellular automaton allowed those cells to receive information from the third dimension, and structures built from such cell states actually become part of this higher dimensional system. This is analogous to the process by which the brain is transformed in the presence of DMT: the brain is restructured at its most fundamental level. All forms are constantly emerging and, in the normal waking state, the brain emerges, from the level of the Grid, from information instantiated by Cell states that receive information only from the limited dimensions of the Grid. However, in the presence of DMT, the brain *re-emerges* from Cell states sensitive to the HyperGrid's orthogonal higher dimensions.

the brain is transformed into a higher-dimensional processor
within the HyperGrid.

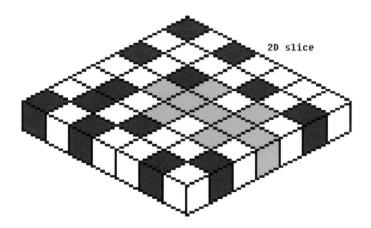

2D slice

A structure emerges as a pattern of blue 3-i state cells
within a 2D slice_

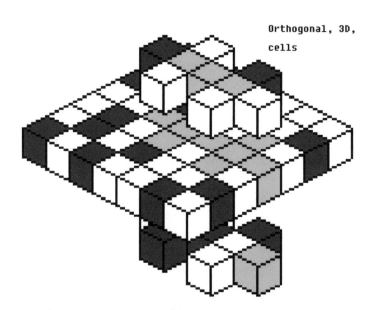

Orthogonal, 3D,
cells

Since 3-i states receive information from 3 dimensions, the
blue structure is revealed to be a 2D slice of a 3D structure_

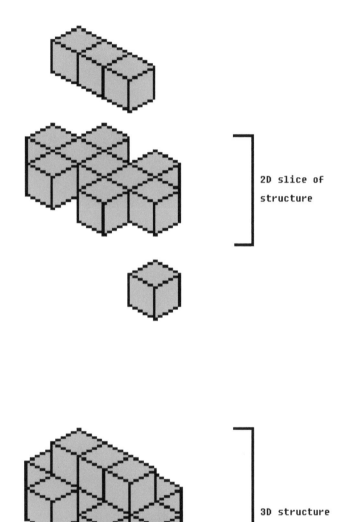

2D slice of
structure

3D structure

Information received from orthogonal
dimensions of the HyperGrid

Information received
from the Grid

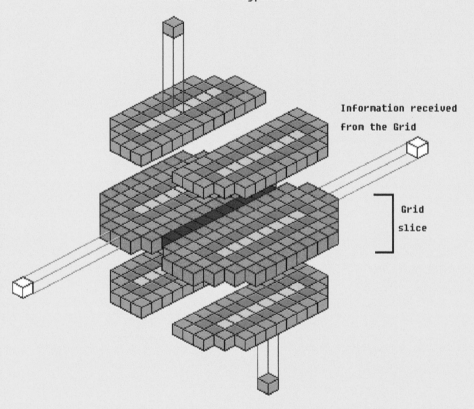

Grid
slice

In the presence of DMT, the brain begins receiving information
from normally inaccessible orthogonal dimensions of the
HyperGrid, revealing itself to be a lower-dimensional slice of
a higher-dimensional information generator and processor_

It might seem difficult to reconcile such a fundamental shift in the structure and behaviour of the brain — from a 3-dimensional to a 4(+)-dimensional structure — with the relatively subtle effects on brain activity observed from the outside. However, it's useful to remind ourselves that the information generated by the brain — by the integrated areas of the cortex forming a T-state — has the special property of *subjectivity*:

the information is experienced as the phenomenal world.

Furthermore, the brain is itself constructed from this information, as well as the information generated at all levels of the organisational hierarchy. As DMT floods the brain, the profound changes that occur in the information it generates are observed from a completely unique subjective perspective.

It's somewhat analogous to the difference between *observing* the patterns of ripples on the surface of water and actually *being* the water. A camera set up above the water will only detect changes in the 2-dimensional patterns of the ripples, but the water — if it was capable of having an experience — would have access to the full dimensionality of its dynamics: the vibrational modes pulsating in the third dimension and, perhaps even, the endlessly dynamic agitations and interactions of its constituent water molecules. Modern functional neuroimaging is the camera observing the limited dimensions of the brain complex from a less than privileged position — from the *without* — whereas the DMT user is observing these processes from the within_

within.   within.

within.   within.

Information is generated at every level of an organisational hierarchy, from the Grid upwards. In the normal waking state, the individual's phenomenal world — the information generated by the brain experienced subjectively — is constructed almost entirely from high-level information generated by global cortical activity, from the integrated activity of large numbers of neurons within an active T-state. The thalamocortical system seems to sit at just the right level of organisation for encoding information about the gross patterns of information that emerge in the environment and thus it makes sense that the phenomenal world is constructed using information generated at this level of the hierarchy. As you scan your environment for potential predators, it isn't necessary for you to be aware of the multitude of intricate molecular interactions inside the cells of the organisms that surround you, but only their high-level structure and behaviours:

Who are they? Where are they? What are they doing?

What is their intent?

This *coarse-grained* cortical level of information overwhelms the phenomenal world and there is no awareness — under normal circumstances — of information being generated at the level of cells, molecules, or lower. However, this lower-level information is also being generated by the same areas of the brain responsible for generating the phenomenal world information, except it is excluded from subjective experience. In fact, this lower-level information is absolutely essential for generating the higher-level information, since the upwards flow of information drives the emergence of the higher levels of order.

However, during the DMT state, when information begins to flow from the normally inaccessible dimensions of the HyperGrid, there is an expansion of this awareness from the cortical level downwards through the hierarchy to the level of the [Hyper]Grid. This is because the phenomenal world of the hyperdimensional brain complex is generated from this lower-level, higher-dimensional, information, in addition to the information generated by the networks of the thalamocortical system. Many DMT users actually experience this expansive downshift during the early phase of the trip as a feeling of being propelled downwards to the core of reality, sometimes observing the transition in stages down through the molecular, atomic, and subatomic, before reaching the breakthrough phase in which hyperspace itself manifests.

Normal waking state	DMT state

Cortical

Intermediate
levels of
organisation

[Hyper]Grid

Awareness

Of course, this is a shift in awareness that is only available to the
individual undergoing the experience: to an outside observer, all that
can be measured is the change in information — neural activity — being
generated at the cortical level, as revealed by modern functional brain
imaging techniques. There is no way for these instruments to detect this
expansion in the dimensionality of the brain complex, but only the effect of this information as it flows upwards to the level of the cortex,
perturbing its global dynamics and modulating the information generated
at that higher level.

When DMT gains entry into the brain, it causes an entirely new brain to
emerge: the brain is transformed from a lower-dimensional processor — an
information complex emergent within our slice of the HyperGrid — to a
hyperdimensional jewel of transcendent beauty and magnificence embedded
in the HyperGrid — but it is a jewel that can

only be enjoyed from within.

>>Chapter 13:
>>The Mechanism of
>>Interdimensional
>>Communication_

"What happens is, the world is completely replaced, instantly, 100 percent. It's all gone. And what's put in its place, not one iota of what's put in its place was taken from this world. So it's a 100 percent reality channel switch."

Terence McKenna

In radiocommunications, the term *channel switch* refers to the shift from receiving an input signal at one particular radio frequency band to another. A simple crystal radio tuner, for example, uses a capacitor to modulate a tuning coil to resonate with, and amplify, a particular frequency of electromagnetic wave whilst ignoring the others. As the tuner is shifted to a new frequency, reception of the original signal is lost as the new one begins to crackle through the speakers. The radio tuner modulates the flow of electricity through the tuning coil so that it can receive the information embedded in the new frequency. Analogously, DMT modulates the flow of information through the brain such that it loses the ability to receive information from the consensus world — from our dimensional slice — but gains the ability to receive information from the normally hidden orthogonal dimensions of the HyperGrid, experienced as the DMT hyperspace. This complete switch of the phenomenal world from *consensus reality* to the DMT *hyperspace* has two specific requirements:

1. Information from the orthogonal dimensions of the HyperGrid must be gated into the brain_

2. This orthogonal information must modulate — by being *matched* to — the ongoing intrinsic information being generated by the brain complex such that the phenomenal DMT world is built.

Remember, the construction of a phenomenal world by the brain is not a passive process: your phenomenal world is always built from intrinsic information, with extrinsic sensory information only modulating this information by being *matched* to it. This applies under all circumstances, including a DMT trip. DMT doesn't transfer your consciousness from the consensus world to the DMT world but, rather, DMT causes your brain to cease building the consensus world and start building the DMT world. But, for the brain to achieve this, it's not enough that it can receive information from these orthogonal dimensions (requirement 1). This information must successfully modulate the intrinsic information generated by the brain complex, which means the intrinsic information must change so that it *matches* this information from the orthogonal dimensions (requirement 2, see page 123). DMT has a crucial role in both of these processes.

During waking life in the consensus world, in-formation enters the brain complex from within our dimensional slice through the normal senso-ry apparatus. This is well understood, although it's helpful to remind ourselves of what's ac-tually going on:

light, sound, and other sensory stimuli are, in fact, patterns of information interacting with another, extremely complex, pattern of informa-tion known as your brain. Your sense organs are highly evolved structures that are optimised for receiving certain patterns of information from this dimensional slice and feeding them into specific areas of your brain complex for further processing.

Obviously, the information received from the orthogonal dimensions of the HyperGrid during a DMT trip is not received through these sensory conduits, and it would be wrong to assume that information transfer into the brain can only occur using the usual sensory apparatus. If I were to strike you hard enough on the head, you might well see flashes of light. This is not from light information, or even stimulation of your retinae, but because the rattling of your brain inside your skull altered the information it generates. Information transfer occurred not via the senses, but by direct stimulation of the brain, albeit in a rather uncontrolled manner. To understand how DMT allows information from orthogonal dimensions to enter the brain, we must let go of the assumption that the familiar sensory routes are the only means of information transfer into the brain complex.

>>>The 3 phases of

>>Entry into the DMT reality
>>("breaking through") can be
>>broken down into three phases:

>>1. Bottom-up modulation
>>[activation phase].
>>
>>
>>2. Top-down modulation
>>[gating phase].
>>
>>
>>3. Transdimensional
>>informational feedback loop
>>[lock phase].
>>
>>
>>
>>
>>
>>
>>
>>_

Activation phase

During normal waking life in the consensus world, your brain generates the intrinsic information that constitutes your phenomenal world. Extrinsic information from outside the brain, received via the senses, modulates this ongoing intrinsic activity such that a stable, predictable, and adaptive phenomenal world is built. When DMT enters the brain, it binds to and activates the 5HT2A receptors embedded in the membrane of the cortical pyramidal cells. The effect of this is to initiate a sequence of reactions and interactions inside the pyramidal cell. Since all of these effects are occurring at the molecular level, this is *sideways information flow*, with the information generated by the DMT molecule binding to the 5HT2A receptor flowing into the network of molecules inside the neuron. However, since this network possesses emergent properties – such as the ability to regulate the membrane potential of the neuron – the information also flows upwards, causing the membrane potential to rise towards the threshold potential. This is the *depolarising* effect of 5HT2A receptor stimulation and occurs in pyramidal cells across the cortex. The pyramidal neurons form part of the highly complex emergent system that generates the patterns of intrinsic information – T-states – that constitute your phenomenal world, and this particular adjustment of their membrane potential by DMT causes that pattern of information to change. Your world begins to change. This is an upwards flow of information from the molecular level to the cortical network and phenomenal world level. For most psychedelic drugs, each with their own effect on the pattern of intrinsic information generated by the cortical network, this is the end of the story. The change in the pattern of intrinsic information is the change in the world generated by the drug: the psychedelic effect. However, with DMT, this is only the beginning. This phase is the initial ACTIVATION of the brain by modulating the information it generates.

Gating phase

All psychedelic drugs affect the intrinsic information generated by the cortex in their own particular way. Usually, the effect can be likened to a kind of 'loosening' of the information such that the phenomenal world becomes more fluid and unpredictable than the normal waking world. This is achieved by expanding the repertoire of T-states to include entirely novel states. DMT, however, exerts a more specific effect: rather than simply rendering the information more random, DMT shifts the intrinsic information from one pattern,

the *consensus world pattern*,
to another pattern:
the *DMT world pattern*.

[see page 123]

This is more analogous to switching a channel to an entirely different frequency than to nudging the dial sightly out of tune. However, this channel switch doesn't explain how information from the orthogonal dimensions can access the brain. To explain this, we must return to the idea of downwards information flow, which occurs when high-level information selects from a lower-level state space.

DMT elicits a highly specific pattern of information from the activity of the cortical system. Assuming the dosage is sufficient to surpass the initially disorderly *activation phase*, the information becomes 'hyper-structured'. The DMT user will notice characteristic DMT patterns and motifs when entering this phase — extremely complex and almost impossible to describe, but unquestionably 'DMT-esque' in their form. This hyper-structured information flows down to the level of the Grid itself, selecting specific patterns of Cells that we have referred to as *perturbation brain patterns* (see chapter 11). These DMT-elicited Cell configurations only emerge within the brain complex in the presence of this particular psychedelic molecule and are mapped to the rule sets that allow 4(+)-i Cell states to emerge. So, overall, the information generated during the activation phase flows to the level of the [Hyper]Grid and gates the flow of information from the orthogonal dimensions of the HyperGrid into the brain complex.

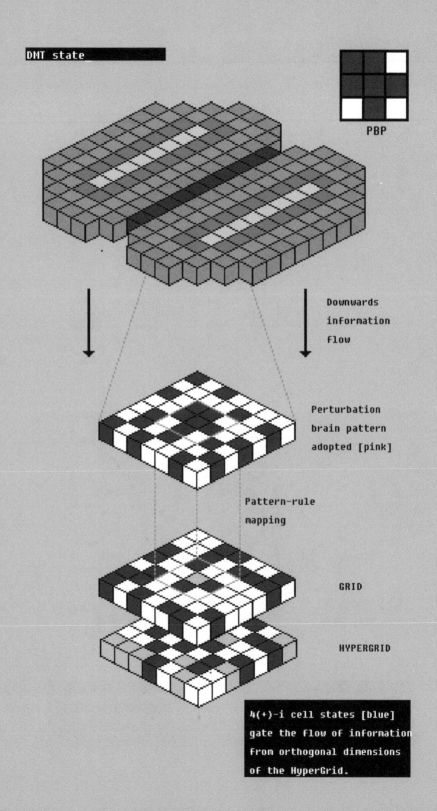

DMT state

PBP

Downwards
information
flow

Perturbation
brain pattern
adopted [pink]

Pattern-rule
mapping

GRID

HYPERGRID

4(+)-i cell states [blue]
gate the flow of information
from orthogonal dimensions
of the HyperGrid.

Intrinsic
information builds
the consensus
phenomenal world

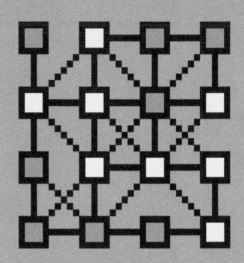

Sensory matching

**Information
from Grid**

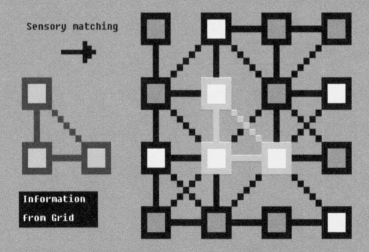

In the normal waking state, the thalamocortical system builds
the consensus world as a model of the Grid, modulated by
extrinsic information from the Grid [sensory matching]_

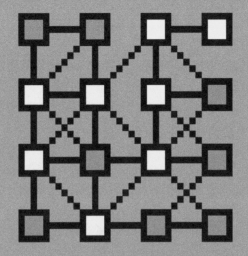

DMT-modulated
intrinsic
information builds
hyperspace

Sensory matching

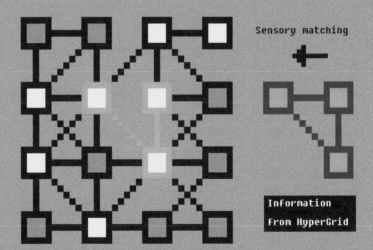

Information
From HyperGrid

DMT perturbs the activity of the thalamocortical system such
that it ceases to build the consensus world, but begins
to build the DMT hyperspace as a model of the HyperGrid,
modulated by information from its orthogonal dimensions_

Lock phase

DMT binding to the 5HT2A receptor doesn't open a magical portal to another dimension, but causes a specific pattern of activity — information — to be generated by the thalamocortical networks. This high-level information then flows downwards to the level of the [Hyper]Grid, selecting Cell state patterns [brain perturbation patterns] that allow information to be received from its orthogonal dimensions.

The information received from the orthogonal dimensions then begins to flow upwards in the usual way through the molecular and neuronal levels, up to the cortical network level, further modulating cortical activity and the information generated by the cortical networks. So, information is flowing in both directions: downwards from the cortical level to the level of the [Hyper]Grid and upwards from the [Hyper]Grid to the cortex. Of course, there is nothing unusual about information flowing in both directions through the layers of a complex system. The difference here is that the DMT-modulated high-level information generated by the cortex has a unique effect at the level of the HyperGrid, owing to the *pattern-rule mappings* which actually gate the flow of information from the HyperGrid.

A brain complex is special in that the complex, emergent information that it generates is highly flexible and dynamic, allowing information transfer from the orthogonal dimensions of the HyperGrid to modulate the intrinsic activity being generated by the brain complex in real time. Of course, this means that the information flowing downwards from the cortical level to the HyperGrid is also altered. The top-down effect of this new modulated information is more efficient at generating the *perturbation patterns* at the level of the HyperGrid to further enhance information transfer from its orthogonal dimensions. This information flows up the hierarchy to the level of the cortex and so on. So, a positive feedback loop emerges, with information from the orthogonal dimensions modulating cortical information such that the transfer of information from those dimensions is enhanced, allowing them to further modulate cortical information. This *information feedback loop* eventually 'locks' the cortex into the state in which it is most efficient at receiving information from the orthogonal dimensions and constructing the phenomenal world experienced as hyperspace.

Prior to this stage, there is an experience of confusion, disorienta-
tion, and a flurry of highly complex imagery and influx of information,
and only when the *lock phase* is achieved does this stabilise. *This is
the breakthrough*. For this phase to be reached, however, the activity
of the cortical system, in the presence of DMT, must be such that the
information flowing upwards to the cortical level can be absorbed by the
cortical networks. Remember, just as normal sensory information can only
be absorbed by the brain if it matches its ongoing intrinsic activity, so
it is with information from the HyperGrid. In the presence of DMT, the
modified intrinsic activity is matched to the information being received
from the HyperGrid [whilst the brain loses the ability to absorb information via the normal sensory apparatus], which can then modulate cortical
activity as the brain begins to build a stable model of the HyperGrid,
which is experienced as the breakthrough into hyperspace.

Crucially, the effect of DMT on cortical activity is determined by its
connectivity and, since thalamocortical connectivity varies between individuals, some people might be more or less sensitive to the transdimensional gating effects of DMT, with a small proportion never progressing
beyond the *activation phase*. However, assuming the *lock phase* is successfully reached and stabilised, the brain complex is no longer restricted
to our usual dimensional slice of the HyperGrid: the brain re-emerges
from the HyperGrid as the hyperdimensional brain complex, and the tripper
becomes part of that higher-dimensional reality.

This is when the elves welcome you home.

It's important to note the two distinct, but overlapping, roles of DMT
in this process:

1. DMT perturbs cortical activity to generate the *perturbation brain
patterns* at the level of the Grid: initial dimensional gating.

2. The DMT-perturbed cortical activity also allows the brain to absorb
the information from the HyperGrid as it flows to the cortical level.

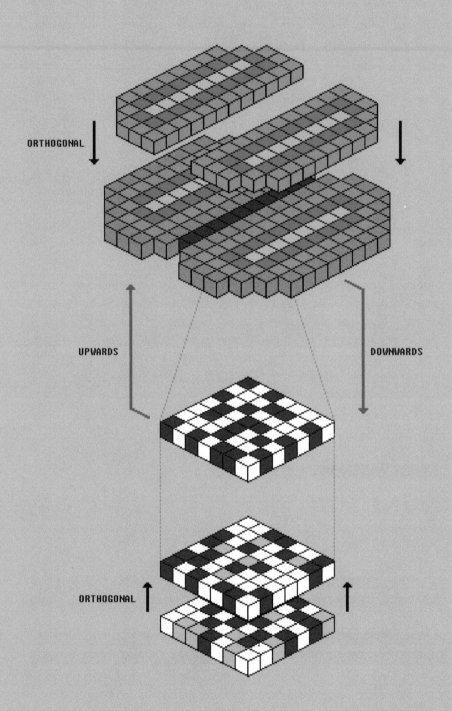

ORTHOGONAL

ORTHOGONAL

UPWARDS

DOWNWARDS

ORTHOGONAL

Unfortunately, [or fortunately], this state doesn't last forever: after only a few minutes, the DMT levels drop below a threshold and serotonin, which is essentially competing with DMT at the receptor site, again floods the 5HT2A receptors. In the presence of serotonin, the pyramidal neurons are returned to their basal activation level and this reverses DMT's effects on cortical activity. The brain begins generating the same patterns of information as before DMT was ingested, and the hyper-structured, downward-flowing information generated by the brain in the presence of DMT is no longer generated in its absence. As such, the positive feedback loop is broken and the brain loses access to the orthogonal dimensions of the HyperGrid. Feedback loops often tend to show switching behaviour, shifting rapidly from an inactive to a fully active state, and vice versa. This is why the transition from the DMT space back to consensus reality is often abrupt. You might feel as if you are suddenly jolted back into the consensus world, shaking, awestruck, and grateful for the most horrifyingly beautiful and astonishing experience you could never have imagined.

Information flowing from the ORTHOGONAL dimensions of the HyperGrid [see gating phase] flows upwards and modulates information at the cortical level. This modulated information flows downwards to the level of the [Hyper]Grid, further enhancing the selection of brain perturbation patterns and establishing a positive feedback loop that locks the brain complex in the DMT space_

Chapter 14:

The Structure

"The universe is information and we are stationary in it,

not three-dimensional and not in space

or time.

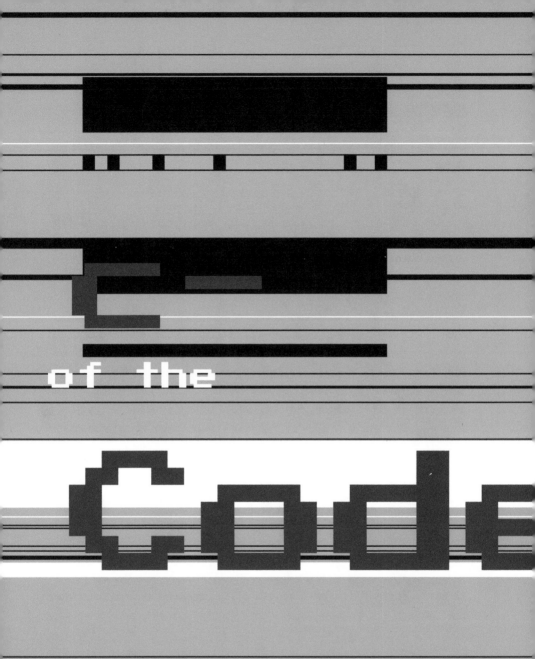

of the

Code

The information fed to us we hypostatize into
the phenomenal world."

Philip K. Dick, VALIS

The Code generates the [Hyper]Grid, which instantiates the
information from which all emerges in this reality:

everything is a manifestation of the
complexification of information.

In our effort to understand the nature of the reality within which we
are embedded, we are now ready to strip away that last layer of con-
ceptualisation to arrive at the true sructure of the Hyper[Grid]: pure
digital information generated by the Code. When discussing the Grid, we
have used cellular automata as exemplifying its essential features: a
digital universe of connected Cells updating their state with each click
of time, based on the states of Cells in their neighbourhood. Further, we
have always assumed, albeit implicitly, the spatial nature of a cellular
automaton grid, whether in two, three, or higher, dimensions. For example, Conway's Game of Life is a 2-dimensional digital 'universe' in which
the infinite spatial extent of the grid seems to be fundamental to the
structure of the automaton: each cell has a particular spatial extent – a
square area of the grid that it occupies – and complex structures emerge
on this grid extending infinitely in the four directions of the plane.
There is a very specific reason for visualising the Game of Life as a
2D square grid: the state of each cell updates depending on the states
of the eight cells in its neighbourhood, making a square grid the most
natural way to represent the relationships between the cells. However,
there is nothing in the construction of the Game of Life that requires it
to be visualised in such a way. An alternative, and much more flexible,
way of representing a cellular automaton is to use a network of NODES
connected by EDGES instead of square cells connected side by side. A *node*
performs the exact same role as a *cell* (in fact, they are fundamentally
the same) and can exist in a finite number of states – two in the case
of the Game of Life – with the edges indicating the connectivity. The
Von Neumann neighbourhood, for example, is a 1:4 pattern of connectivity
– the central cell is connected to four surrounding cells – whereas the
Moore neighbourhood is 1:8.

The standard grid representation of a cellular automaton doesn't
indicate the neighbourhood used in the transition rules, whereas this
is made explicit in the equivalent network representation_

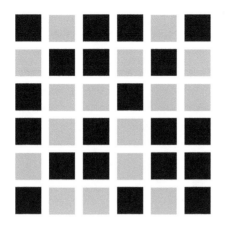

Standard grid
representation

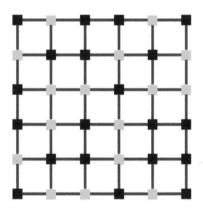

Node representation
[Von Neumann n'hood]

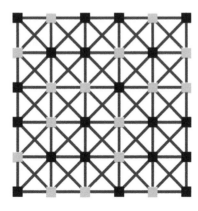

Node representation
[Moore n'hood]

Although certain ways of representing a cellular automaton might be preferred over others, there is actually no requirement for the automaton to be visualised at all — that is purely for our enjoyment. Each cell of the Game of Life, for example, is not actually a square area of a grid, but simply a piece of information instantiated by some abstract ELEMENT that can switch between one of two states. In most cases, this element is a component of the computer chip responsible for storing the states of all the cells as the game runs. But coins, checkers pieces, or squares of coloured paper — or any *things* that can exist in precisely two states and thus each instantiate a single *bit* of information — are equally valid, albeit rather impractical, substitutes for the internal states of a computer microprocessor. The material used to represent the states of the cells is completely independent of the patterns of information generated by Life as it runs.

Chess is another game played on a square grid: the pieces on a chessboard might be elaborately carved from exotic hardwoods and are given fancy monikers like *knight*, *bishop*, or *queen*. Or, a modern game of chess might be played entirely on an electronic device. Advanced players might even play out an entire game without ever looking at a board, but simply by recording their moves in algebraic chess notation. At its most stripped down level, the game of chess seems to be something much more abstract and fundamental than knights, bishops, and wooden boards — the game of chess is a MATHEMATICAL STRUCTURE, which we can define, informally, as:

a set of elements with certain defined properties and relationships between each other.

The term *element* is deliberately abstract, in being *a mathematical object with no physical or other types of characteristics other than those specifically given to it*. Think of an element as an imaginary hook upon which we can hang whatever features we want or require. Each *piece* in a game of chess is an element upon which certain characteristics are defined: how does the element move around the space defined by the board? How does the element interact with other elements? Once the specific characteristics of each element are defined, we can play chess on a board, a computer, or by mailing our moves scribbled in chess notation back and forth across the world.The game, at its most fundamental, is independent of how we choose to represent it.

Similarly, at its most fundamental level, the Game of Life is a mathematical structure consisting of a large number of elements together with a set of defined relationships between them. Each element is given two discrete states, only one of which can be occupied at any time, and the ability to receive information about the states of a well-defined selection of other elements — the neighbourhood — and *nothing more*. As Life runs, information is generated as every element updates its state in parallel, selecting from the two possible states based on its current state and those of its neighbours. For ease of visualisation, we have used square *cells* to represent the elements and their states, but

any set of objects — whether concrete or purely abstract — that preserves the relationships between the elements is a perfectly equivalent representation of the Game of Life.

So, cells on a square grid are no more a fundamental component of the Game of Life than bishops hand-carved from Indian rosewood are a fundamental component of the game of chess. Our choice of representing the Game of Life is just that: a representation of something more abstract and fundamental. The essential features of the Game of Life are the possible states of each element, the way these elements are connected (1:8), and the update rules. None of these features require any sort of grid at all, as long as the defined relationships between the elements are maintained.

We think of the Game of Life, in its original form, as being 2-dimensional because of the way it is usually visualised using a 2-dimensional square grid. However, of course, Life can be run on a computer without ever plotting the output as a grid, network, or any other representation. In fact, we can choose to output the information generated by the Game of Life as binary code, as patterns of light or sound, or not at all. How we choose to represent the information generated by Life has no bearing on the information itself, other than perhaps making it easier for us to interpret this information: by observing patterns on a grid rather than trying to make sense of a string of binary digits. This naturally raises the question as to whether the Game of Life remains a 2-dimensional cellular automaton even if its output isn't visualised on a 2-dimensional grid, if at all. To answer this question, we need to think a little more deeply about what we mean by 'space' and the varying dimensionalities that it can possess.

The awareness of a 3-dimensional spatial world is one the most fundamental features of our existence, established from the earliest stages of our lives as we explored our environment and each developed our own unique phenomenal model of the world. Space seems absolutely fundamental to our existence and it feels natural to assume that the ground of reality must have a spatial aspect. We intuitively think about space in terms of the relationships between objects *within space* rather than trying to envisage space itself. Space appears to be an empty container within which objects can be placed. However, space itself has a structure that emerges from something more fundamental. Like all things, barring the Code itself, space is emergent. Mathematically, space is defined as a set of POINTS together with a set of relationships between the points known as a TOPOLOGY.

Most of us are familiar with what is known as the EUCLIDEAN or STANDARD TOPOLOGY which, in its 1-dimensional form — 1-space — is a simple line formed from an infinite number of points, like beads on a thread each with precisely two neighbours (as with a cellular automaton, the points surrounding a point form its neighbourhood):

1-space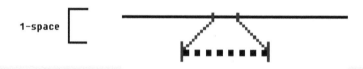

Generating 2-dimensional space — 2-space — is as straightforward as extending the 1-dimensional line by adding a second, orthogonal, line:

2-space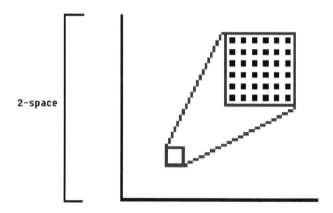

And, adding a third line generates the 3-dimensional space that feels so natural and obvious when we think of space. Using the 1-dimensional standard topology to build increasingly higher dimensional topologies is a common way of building such spaces in mathematics, and there is nothing stopping us adding a fourth, fifth, or even more dimensions to the familiar standard 3-space, by adding more and more orthogonal 1D lines. Although such dimensionality is difficult to visualise, there is nothing mathematically unusual about extending Euclidean space in this way. The dimensionality of a space simply corresponds to the number of independent coordinates required to specify the position of a point in that space. The topology of a space describes the connectivity between those points or, equivalently, the topology defines the shape of the space.

The *standard topology* is the easiest to visualise, extending infinitely and uniformly along each of its dimensions. But an infinite range of spatial topologies can be constructed, from the familiar to the exotic, unexpected, or unimaginable. For example, one can take a finite square of the standard 2-space topology — a simple 2D plane — roll it into a cylinder and then bend the cylinder round to join end-to-end. This creates a torus, or doughnut, which defines a 2-dimensional topology in many ways similar to the standard 2-space topology, except the relationships between the points on the topology are different: In the standard topology, the points extend infinitely in each of the two dimensions but, on a torus, moving along either dimension will eventually bring you back to where you started. The same can be said for the surface of a sphere, another very familiar 2D topology. Topologies can be defined in any finite, or infinite, dimensional space, but many of these can only be grasped in complete abstraction. The important idea is that the relationships between the points of a space, whatever its dimensionality, are defined mathematically and can be encoded. You don't need a doughnut to define a doughnut topology. And, of course, a cellular automaton is simply a set of elements — or points — together with the set of defined relationships between them. In other words, a cellular automaton defines a topology, a space. The dimensionality of an automaton's topology is not dependent on how it is represented, on a grid or not at all, but is intrinsically defined by its connectivity. So, the Game of Life is 2-dimensional because of the connectivity of its encoded elements (cells/nodes), not because of the conventional choice to represent it on a square grid.

Each cell of this 4x4 grid can be fully represented by three integers: The first two represent its position on the grid [one for each dimension], and thus its relationship to its neighbours, and the third represents its state [0 or 1]_

The entire grid can thus be represented as a string of 48 integers:

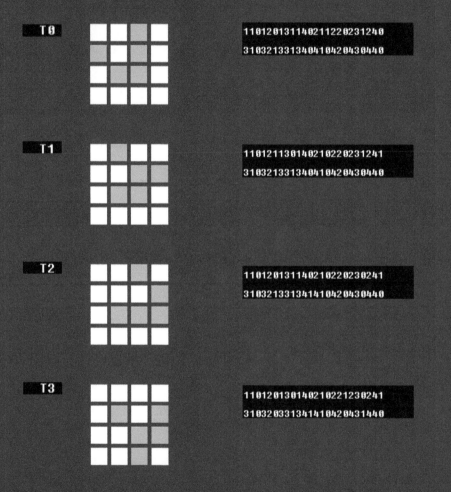

T0
110120131140211220231240
310321331340410420430440

T1
110121130140210220231241
310321331340410420430440

T2
110120131140210220230241
310321331341410420430440

T3
110120130140210221230241
310320331341410420431440

Only the grid representation clearly reveals a dynamic pattern of information we know as a glider. However, both representations are entirely equivalent and contain the same information_

The properties and information patterns generated by the Game of Life, including the dimensionality and topology of its space, are completely independent of how, or whether, we might choose to display this information: as a 2D grid, network of nodes, or as a binary string. A 2D grid is merely a convenient way for us to visualise the information generated by Life from the *outside*. It's important that we don't erroneously think of the spatial topology of Life as somehow being a *representation* of space, in the way that an architect's computer model of a house is a representation of its 3-dimensional structure. In a cellular automaton, the topology encoded by the relationships between its cells *is* the actual space. To appreciate this, we can reposition ourselves, not as outside observers, but as complex critters that emerge *inside* the Game of Life.

Assuming that conscious beings
conscious beings
conscious beings
conscious beings
conscious beings

conscious beings

might eventually emerge in a cellular automaton running on a computer, as one of these emergent critters you would experience a reality with a

dimensionality commensurate with the topology defined by the automaton:

The dimensionality and spatial quality of your phenomenal world would be determined by the connectivity of the cells as the game runs. Or equivalently, the connectivity of the cells would define a topology that you would experience as your *space*, in the same way we experience our usual 3-dimensional world outside of the game. Looking around, you would notice that each *point* in the space was connected to eight other points and that travelling between points required you to cross a specific number of other points. Whether or not the output of the automaton was displayed on a convenient screen for the viewing pleasure of humans *outside* the game would be entirely irrelevant to your experience as a conscious being *inside* the game.

Of course, this also applies to automata of higher dimensionality:

> a conscious critter that emerges inside a 9-dimensional cellular automaton would experience a type of 9-dimensional space that we would have no way of ever comprehending without entering it.

And, likewise, this applies to our reality: although the discrete points that comprise our reality — the Grid — are too small for us to observe directly, we appreciate and experience the gross features and qualities of our 3-dimensional spatial topology. Some of these features we can point out and measure, such as the three orthogonal directions of space and the distances between objects. However, our experience of 3-dimensional space is largely ineffable, like the quality of the colour red, and can only be understood by being experienced. This is why, no matter how intellectually or mathematically prepared one might be, tumbling into DMT hyperspace space is thoroughly confounding — there is no way to anticipate the quality of a such a space before one enters it. Nor indeed is it easy to render the space in the memory once one returns, other than perhaps as a 3-dimensional projection or approximation, which perhaps partly explains the frustration commonly experienced in trying to recall details of the DMT state.

So, the Code is a structure of unknown (and irrelevant) dimensionality that encodes the finite states of the discrete ELEMENTS/NODES/CELLS of the Grid, together with the connectivity between them, in much the same manner that the binary code inside a computer generates, stores, and updates the states of the cells in the Game of Life. *There is no fundamental spatial aspect to the Grid: the connectivity between the elements/nodes/cells is encoded by the fundamental rule set and defines the dimensionality and topology of the space that emerges as the Code generates the Grid and, ultimately, defines the quality of the space experienced by self-reflective conscious organisms that eventually emerge within the Grid, including ourselves.*

Since an alien hyperintelligence was responsible for generating the Code, and since the Code generates the [Hyper]Grid, it is tempting to conclude that we are the products of some kind of intelligent design. But this is not the case, for the Code doesn't only generate the Grid upon which we find ourselves, but *all* possible Grids.

Chapter 15:

HOW TO BUILD A UNIVERSE

Our Universe is a lower dimensional slice — the Grid — of a higher dimensional system — the HyperGrid, which is generated by a fundamental Code. Although the Code was generated by an alien hyperintelligence outside the HyperGrid, we were not designed. When John Conway originally encoded the Game of Life, he had no idea that the program, when run, would exhibit such rich complexity — he happened to stumble upon one of the few rule sets that spawn a cellular automaton with interesting behaviour. For any particular type of automaton, there is a finite — albeit potentially extremely large — number of possible rule sets, which form the *rule space*. Every rule set within the rule space will display behaviour somewhere along the continuum from *Type I* homogeneity to *Type III* chaos, with just a few sitting in that narrow — *Type IV* — band between order and chaos, where complex, stable structures emerge.

So, how does one go about finding a cellular automaton with complex, open-ended, behaviour that might eventually harbour living organisms? There seems to be no simple way of knowing whether any particular rule set will engender the type of structures with the potential to complexify towards conscious life: running the program is the only sure test. Building a universe within which complex life will emerge is as easy as finding the right rule set. And as hard.

Cellular automata exemplify the idea that simple code, when executed, can generate behaviour that is startlingly complex and thoroughly unpredictable. This is the Game of Life written in the ultra-concise APL language:

```
life←↑1 ⍵∨.∧3 4=+/,¯1 0 1∘.⊖¯1 0 1∘.⌽⊂⍵
```

Writing the code to enumerate all possible rule sets — to generate every rule set within a rule space — for a cellular automaton is fairly straightforward. If we take a 1-dimensional elementary cellular automaton, for example, there are precisely 256 rule sets to iterate through. As the size of the neighbourhood and dimensionality of the automaton increases, the size of the rule space expands rapidly. But, crucially, however large it may grow, it will always remain finite and, with sufficient computing power, the entire rule space can be explored exhaustively, even for the type of system that forms the basis of our Universe (Grid).

A larger rule space doesn't mean that more complex code is required to enumerate through it, only that the output and the time and requisite computational resources will be much larger. Crucially, it will always be easier, and computationally cheaper, to run every possible automaton — every possible Grid — than to attempt to determine *a priori* the rule sets that will produce the type of behaviour one hopes will emerge[1].

However, it isn't necessary to run every rule set until it either does or doesn't yield living organisms. Experience will indicate that certain patterns of behaviour in an early Grid — universe — tend to augur the eventual emergence of complex life, whereas other patterns preclude it. Grids with such potential can be selected, whilst those that fail to display such behaviour are halted. GAME OVER. Naturally, a great deal of intelligence and experience will be required to distinguish Grids with such potential from those destined for either endless homogeneity or a ceaseless tumult of structureless chaos. However, the decision to either keep a Grid running or to halt it can usually be made early.

Once a universe emerges with apparently intelligent, conscious life, it's natural to wonder what one might do with such a universe or indeed with the conscious intelligences emergent within it. Broadly, there are two options:

> leave the universe alone, do not intervene or interfere in any
>> way, or;
> intervene in or interact with the universe in some way towards
>> some end.

If the former approach is taken, the intelligences within the universe will presumably forever remain ignorant of their status as part of an encoded, emergent, reality. However, it's possible that a sufficiently advanced intelligence might eventually reach a level of cognitive and scientific sophistication such that their status is suspected and, potentially, tested. Indeed, within our civilisation there already exists a clutch of plucky philosophers and physicists that have proposed ways of testing whether or not we live in a 'simulated reality'[2]. This terminology is avoided here, since it presumes that a reality constructed from digital information must be simulated, whereas, in fact, our Universe is an *instantiation* of a reality rather than a simulation of one.

In the case of our Grid, its ultimate purpose is the Game, which involves conscious intelligences interacting with — interfacing with — the HyperGrid. We will deal with the detailed nature of the Game, and how we can play it, in the final chapter. For now, having developed a deep understanding of the structure of our reality — the Grid — and how it was generated from the Code, together with what we have learned about how the flow of information between dimensions of the HyperGrid is controlled, we can begin to understand how and why DMT has the special property of gating information flow from the HyperGrid's normally hidden dimensions into the human brain.

Although there are certain universal features of conscious self-reflective information complexes — brains — that emerge on our Grid, there is no simple way to 'plug' such a complex into the HyperGrid. Imagine programming and running a simulated world on a computer sitting on your desk. To your astonishment, complex living forms begin to emerge within the digital world, living beings that, eventually, begin displaying signs of conscious intelligence. After some time, you decide it's time to try and communicate with these beings, perhaps even attempt to connect them to our 3D world so they can experience our reality. *How would you go about this*?

This is not a trivial problem, since their lower-dimensional world is completely alien to our higher dimensional *container reality*. You might, perhaps, try and identify their brain complex and input information. But, even if you were successful in injecting information into their brain complex, it's certain that it would make no sense to them whatsoever. Unlike such a computer simulation, the structure of the Grid is such that it is set up, from its inception, to receive information from the higher-dimensional reality. Being a dimensional slice of the HyperGrid, it is already fundamentally connected to it in the same way that a 2D slice is connected to a 3D cubic cellular automaton. However, the Grid is informationally isolated from the HyperGrid, since its Cell states are insensitive to dimensions beyond the three dimensions of the slice.

Although, from our perspective, the Grid is isolated from the HyperGrid, in that no information flows from the HyperGrid to the Grid, all of the information generated by the Grid — and indeed the HyperGrid — is available to the

alien hyperintelligence

that generated

the Code.

This means the Grid can be analysed for the characteristically complex patterns of information generated at the level of the Grid by emergent brain complexes: *native brain patterns* (see chapter 11). Once these patterns, and thus potentially conscious intelligences, are identified, the task is to generate the *pattern-rule mappings* that will gate the flow of information from the HyperGrid into the brain. However, as explained in chapter 11, *native brain patterns* are not used for these mappings, since they would result in hyperdimensional information flooding the brain at all times. Rather, modified patterns of activity — *perturbation brain patterns* — are employed. These are highly specific patterns of information that result from the modulation of brain activity by a very particular external information complex which, in our case, takes the form of_

When DMT floods the brain, the patterns of information it generates are dramatically altered and, as this modified information flows downwards to the level of the Grid, it is the altered Cell/Node patterns that are mapped to the rule set that allows 4(+)-i states to emerge, thus gating the flow of information from the HyperGrid into the brain.

1001000000011000010010000000110110101100001011011100110100101100110011001010111001101111010001
0001101101111011011010111000000110110001100101011110000011010010110011001101001011000110110011
0011000011000010111010000110100101101111011011100010000000110111101100110001000000011010010110
1110110110110001000000011011111011001100010000000110010110110110011001100110111101111001001101101
1100000011010010110111100110011011011111011100011011101101011000010111010000110100101101111011

It is natural to wonder why DMT — and DMT only — has this special role as the information gatekeeper, and this will become clear when we discuss the nature of the Game in detail in the final chapter. At this point, it is sufficient to understand that DMT is, in a sense, an intelligence test: whilst it would be straightforward for the author of the Code to map our native brain patterns to the appropriate rules and plug us into the HyperGrid at will, it is an important part of the Game that we plug *ourselves* into the HyperGrid. DMT is the plug, the channel switch, that must not only be identified as such, but must be also be isolated — or synthesised — in a reasonably pure form and the most effective mode of administration found. This requires considerable intelligence and could not be achieved by a creature with only a primitive intellect. No, understanding and using DMT as a technology — and a technology it is — requires a level of cognitive sophistication that has, so far, found an Earthly expression only in humans. Elephants, monkeys, and other beasts might well chew upon psychoactive leaves or munch wind-fallen and sun-fermented berries, but only humans possess the intelligence, and the technical acumen, to use DMT.

On Earth at least, so far, only the human can pass the test.

The Code

Implementation
of the Code

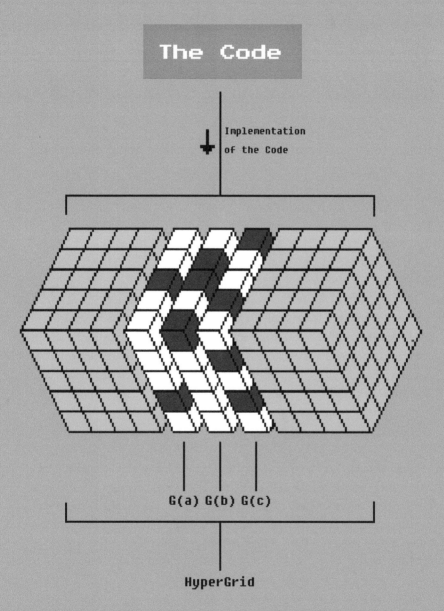

G(a) G(b) G(c)

HyperGrid

Running the Code generates the higher-dimensional HyperGrid and a vast number of lower-dimensional Grid slices (only 3 are shown), within one of which we find ourselves emergent_

Before discussing the nature of the Game and its resolution
discussed so far. The preceding chapters were very much a
is prerequisite:

We live in a reality constructed from digital information instantiated
by a Code programmed by an alien hyperintelligence [the Other]_

The Code programs the structure of our reality as a digital object
constructed as one of a vast number of lower-dimensional slices —
Grids — of a structure known as the HyperGrid_

The Grid — and the HyperGrid — is encoded as a network of fundamental
Cells/Nodes connected according to a rule set. This connectivity allows
information to flow between Cells/Nodes and throughout the [Hyper]Grid.
However, the flow of information between the Grid and the HyperGrid
is normally restricted using 3-i Cell/Node states insensitive to the
orthogonal dimensions of the HyperGrid.

The Cells/Nodes, with their connectivity, form a spatial topology that
determines the dimensionality of the space that emerges as the Grid
runs. Being within a lower dimensional slice of the HyperGrid, emergent
organisms, such as ourselves, experience a space with a commensurately
lower dimensionality than beings emergent in the HyperGrid.

Even the most complex structures that appear in our world (and others)
can be seen to emerge through many layers of hierarchical complexity,
from the [Hyper]Grid to the subatomic to the organismic level. This
includes the brain complex, which is built from information and is an
extremely complex information generator. Everything is a manifestation
of the complexificaton of information.

Information is dynamic and flows through the HyperGrid, both within organisational layers and between them. The upwards and downwards flow of information is crucial for the self-organisation of complex, including living, systems.

As brain complexes emerge and evolve, they eventually construct a model of reality experienced as a phenomenal world. This subjective world is built from vast amounts of information, which the brain generates by selecting T-states from an astronomical number of potential states (and thus ruling out extremely large numbers of states).

Psychedelic molecules modify the information generated by a brain, modifying the structure and dynamics of its phenomenal world.

The key requirement for transport to an alternate orthogonal dimension of the HyperGrid is the transfer of information from normally hidden dimensions into the brain complex. This is achieved using *pattern-rule mappings* that map the *perturbation patterns* generated by the brain in the presence of DMT to rules that engender the emergence of Cell/Node states sensitive to these higher dimensions that gate the flow of information from the HyperGrid into the brain.

Once information begins flowing from orthogonal dimensions of the HyperGrid into the brain complex, upwards and downwards information flow establishes a feedback loop that locks the brain complex in the higher dimensional state, building hyperspace, until DMT is removed.

"Maybe this is not actually a reality. We're trapped, or I'm trapped, I don't know if you're trapped, but we're in some kind of piece of fiction. It's like a Philip K. Dick deal, we're in some kind of simulacrum."

Terence McKenna

Chapter 16:

>> THE GAME

The task of the Game is to realise the nature of our imprisonment in the Grid and to find our way out. But a game it is.

The Game has six levels, the final being *resolution* which involves the transcription of the brain complex and permanent — *irreversible* — transference into the HyperGrid:

>>**Level I: Information**

>>**Level II: Emergence**

>>**Level III: Transmission**

>>**Level IV: Immersion**

>>**Level V: Realisation**

>>**Level VI: Resolution (transcription)**

Levels I and II — Information and Emergence

These first two levels are simply the process by which the Grid engenders conscious intelligences: the Code generates countless variations of the Grid, only a handful of which will result in the emergence of complex, intelligent beings. These Grids evolve independently of the HyperGrid according to their own particular variant of the fundamental rule set dictated by the Code. Of all these Grid variants, most either grind down to endless homogeneity or convulse towards a maelstrom of ceaseless chaos. A few, however, self-complexify towards the emergence of complex, living organisms and, ultimately, to conscious intelligences. However, no matter how complex an organism might become, without intervention, it will always remain independent of the HyperGrid, embedded in its isolated dimensional slice: the Grid upon which it emerged.

The emergence of conscious intelligences within a Grid can be detected by the specific, highly complex, patterns of information generated by brain complexes. However, such patterns do not establish the level of intelligence required for transference into the HyperGrid. Once potential conscious intelligences are detected — and selected as potential candidates for the Game — a technology must be embedded in the Grid that will, not only, gate the flow of information from the HyperGrid into the brain complex, but will also provide a test of sophisticated, high-level, intelligence. *This technology is DMT.*

Level III — Transmission

The primary role of the DMT technology is to elicit highly characteristic patterns of brain activity that will cause the adoption of specific Cell/Node patterns at the level of the Grid. These *perturbation brain patterns* can be used to generate the *pattern-rule mappings* that will engender the emergence of $4(+)-i$ Cell/Node states inside the brain complex and thus gate the flow of information from orthogonal dimensions of the HyperGrid into the brain. However, this interdimensional information flow into the brain complex is not sufficient for entry into hyperspace. Recall that switching from consensus reality to the DMT reality requires, not only, the gating of information from those orthogonal dimensions, but also that this information is matched to the ongoing intrinsic information being generated by the brain. So, DMT triggers the initial change in brain activity that gates transdimensional information flow, but this modified activity must be structured such that it can be further modulated by the information received from these orthogonal dimensions. And, as with sensory information received via the usual sensory apparatus, it is the connectivity of the brain which must be sculpted to absorb the extra-dimensional information as it flows upwards from the level of the HyperGrid. In other words, just as with the consensus world, the brain must learn to build a model of the HyperGrid, such that, in the presence of DMT, the activity of the thalamocortical system matches the information being received from the HyperGrid and the brain builds a model of its structure experienced as hyperspace. This is achieved using a *transmission 'priming' phase* of the Game, which moulds the connectivity of the thalamocortical system and, without which, the *lock phase* of the DMT breakthrough process (see chapter 13) could never be achieved.

During this transmission phase, high levels of DMT are present in the brain, which receives a stream of information from the HyperGrid and essentially learns to construct a model of its structure (see chapter 7 for details on the mechanism). Unless this priming *transmission phase* is completed, even though information might be gated into the brain complex in the presence of DMT – owing to the encoded *pattern rule mappings* – the higher-dimensional information will fail to modulate cortical activity to achieve the lock phase of the DMT breakthrough process. You can also think of this as a kind of *tuning phase*: the initial *pattern-rule mappings* gate information into the brain complex whenever DMT is present, but this stream of information from the HyperGrid gradually modifies the connectivity of the thalamocortical system, further enhancing the brain's ability to absorb information from the orthogonal dimensions. In chapter 5, we described this as an increase in the mutual information between the brain and the environment, and the same applies in the presence of DMT, except the increase in mutual information is between the brain and the orthogonal dimensions of the HyperGrid.

The transmission phase occurs in the prenatal brain, as the foetus is developing inside the womb: DMT levels are markedly and consistency high in the foetal brain, but begin to fall at birth until, only a few months into the postnatal life, DMT levels drop to trace levels[1]. Although explicit memories aren't laid down during this time, the sense of déjà vu that accompanies many DMT trips is likely a result of these prenatal sojourns in hyperspace. This is also why the elves often welcome the tripper back with great celebratory uproar and, indeed, there is cause for celebration, since the next tentative step towards the *resolution phase* has been taken. Many DMT users are struck by the unshakeable feeling that their DMT trips carry them back to where they resided before they were born. And, indeed, soon after the development of the brain inside the womb, DMT levels rise and the transmission phase begins. So, the prenatal brain complex begins receiving information from the orthogonal dimensions of the HyperGrid throughout prenatal development. It is only after birth that DMT levels in the brain begin to wane, serotonin regains full control, and the child becomes fully immersed in the consensus world, receiving information only from the dimensionally-limited Grid. So, newborn children are ferried from a prenatal hyperspace to the lower-dimensional life they will gradually become immersed within, to lose any explicit memory of the luminal realms from whence they came.

Level IV - Immersion

The *immersion phase*, like the subsequent *realisation phase*, can be seen to occur at both the individual level over a lifetime, from birth until death [or resolution], and at the societal level, over evolutionary epochs. Until a species reaches a level of cognitive and technological advancement such that the DMT technology is discovered and developed — the phase we are in now — all humans are destined for full immersion in the Grid from birth until death. This is an important phase, since it allows the brain complex to evolve and mature, and for the requisite intellectual and technical capacities to fully develop before *realisation* is possible. It should also be pointed out that these phases run somewhat in parallel: the *transmission phase* is still occurring during the *immersion phase*, whereas the former occurs on the prenatal side of life, and the latter only after birth. At the species level, *immersion* can be seen simply as the gradual evolution of the human brain complex over many millennia, embedded — *immersed* — in the Grid, isolated from the HyperGrid. In a sense, all beings with some level of intelligence and conscious self-awareness are in an immersion phase, in that they are immersed in the Grid, but the vast majority will never progress beyond it. And, although humans separated from other apes a few tens of millennia ago, we have only very recently transcended the invisible line separating those ever-destined to wriggle in the dust from those with a shot at hyperspace.

Level V - Realisation

The DMT technology only takes up its role as an 'intelligence test' at Level V — *realisation* — at which point it relies on human intelligence for its administration. Realisation is both a process of realising the nature of the Game and of finding a way out of it, using the embedded DMT technology. It is a fundamentally requisite feature of the technology that its administration cannot be accidental or the result of normal feeding or exploration of the environment. Rather, it must be sought, identified, and a proper technique for its usage developed — requiring of a highly sophisticated intelligence.

As with the previous phase, the *realisation phase* occurs at both the individual and societo-cultural level since, whilst the individual is capable of achieving realisation of his nature as being imprisoned in the Grid, the process of discovering and utilising the technology for transcending it occurs over generational timescales. We can observe this process in our civilisation: from the discovery of DMT as the active component of indigenous Amazonian medicines, to its synthesis and purification and, most recently, through the range of techniques for its efficient administration that have been developed, largely by a spirited and enterprising clandestine psychedelic chemistry community. However, realisation also occurs in the individual, often within seconds of the first-ever DMT hit, as an earth-shattering *kensho*-like experience of epochal proportions, both inexpressible and undeniable, often irretrievable but almost always irreversible. In terms of the requisite intelligence, the gulf between the casual consumption of a natural psychedelic — such as *Psilocybe* mushrooms — during feeding and the deliberate administration, using specialised techniques, of highly purified DMT is vast and can only be traversed by an intelligence of prodigious dimensions and depth. Of course, as bearers of such intelligence, it's somewhat difficult to appreciate just how remarkable is the gelatinous structure that has emerged within this particular hominid's skull.

The human species has undoubtedly progressed to this *realisation* phase, within which we continue to advance, albeit in fits and starts, and suppressed by a powerful but ignorant and benighted few. The psychedelic effects of DMT were discovered in 1956, synthesised by Hungarian physician and chemist, Stephen Szára, who was seeking support for his hypothesis that DMT, and not bufotenine, was responsible for the psychoactive effects of the Amazonian *cohoba* snuff[2]. It wasn't until the invention of chromatography at the beginning of the 20th century that isolation and purification of DMT from plant materials became possible. Synthetic organic chemistry was similarly in its infancy at this time, and the first synthesis of DMT wasn't published until 1931. Although a variety of modern synthetic routes to DMT have since been developed, including the Speeter-Anthony synthesis used by Szára, the very first was invented less than a century ago. Whether extracted from *cohoba* seeds or synthesised from simpler precursors, before the dawn of the 20th century, humans were literally incapable of progressing beyond the *immersion phase* of the Game, demonstrating the level of intelligence required.

Having synthesised ten grams of DMT tartrate, Szára's initial experiments ingesting the drug orally were unsuccessful: he swallowed increasing doses of the drug over several days, up to around three quarters of a gram — a massive dose — but with no effects whatsoever. Ready to give up on his ostensibly flawed hypothesis, a colleague suggested he try injecting it:

"In April of 1956, I tested three doses intramuscularly, paced at least two days apart to allow the drug to clear my body. The first dose (30 mg, around 0.4 mg/kg) elicited some mild symptoms - dilation of the pupils and some coloured geometric forms with closed eyes were already recognizable. Encouraged by these results, I decided to take a larger dose (75 mg, around 1.0 mg/kg), also intramuscularly. Within three minutes the symptoms started, both the autonomic (tingling, trembling, slight nausea, increased blood pressure and pulse rate) and the perceptual symptoms, such as brilliantly coloured oriental motifs and, later, wonderful scenes altering very rapidly."[3]

Szára had discovered the secret, not only to the *cohoba* snuff's magic, but also — unbeknownst to him at the time — the secret to exiting the Game he didn't even know he was playing. Szára also unwittingly took the first step in developing DMT as a technology: recognising that DMT is inactive when ingested orally, but must be injected or, as would be discovered in the following decade, vaporised to unmask its extraordinary effects on consciousness. Owing to its simplicity, vaporisation of freebase DMT, usually extracted from the root bark of the Brazilian perennial shrub *Jurema Preta* — *Mimosa hostilis* — remains the most popular mode of administration. For many, there is an attractive romanticism attached to the hand-blown glass pipe, hand-woven rugs, incense, and the other accoutrements of the modern psychonaut. But vaporisation is merely a means of pushing open the doorway, peering through, and gasping at the frightening other beyond the threshold before the door is sharply pressed shut again. Developing DMT as a technology for playing the Game demands we bring our best tools to the table.

Unlike the other classic psychedelics — psilocybin, LSD, mescaline — and their synthetic derivatives, DMT is unique in possessing a number of pharmacological peculiarities that mark it out as being special. Most notably, of course, is the extremely rapid onset and brevity of its effects: following intravenous injection of the drug — the most efficient mode of administration — the effects are noted almost immediately and full breakthrough into the DMT hyperspace occurs within 60 seconds. The voyager will then remain in this space for around five minutes before being jolted back into the consensus world with only residual effects. 20-30 minutes later, the effects are fully resolved. Many are frustrated by this brevity,

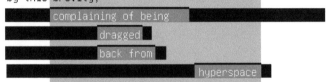

complaining of being
dragged
back from
hyperspace

just when the maelstrom was beginning to stabilise and the entities beginning to speak, although, for most, five minutes is quite long enough. But, for those wanting to return to the DMT worlds in short order, DMT has an additional unique characteristic: *lack of subjective tolerance*. Again, unlike the other classic psychedelics, which display diminishing effects with closely spaced doses, DMT can be injected repeatedly without any loss of subjective potency[4]. Regular LSD users typically abstain from the drug for at least a couple of weeks following a trip. Whilst this is partly as a means of allowing integration of the experience and avoiding negative after-effects, attempting a repeat trip the following day is likely to fail: LSD exhibits rapid and sustained subjective tolerance. Of course, since LSD binds tightly to the 5HT2A receptor, the experience itself lasts for several hours and the 'comedown' phase is sustained. With DMT, in stark contrast, the entry and exit from the DMT hyperspace is clean and rapid and, as soon as the tripper returns, re-entry is as simple as injecting a repeat dose.

The aim of the DMT technology — and required for the *resolution phase* to be completed — is not to provide a brief, jolting, sojourn in hyperspace. Rather, the goal is to establish extended and stable entry into the DMT space: a stabilised *lock phase* of the DMT breakthrough process. The unique pharmacological characteristics of DMT — rapid onset and clearance, brief effect, and lack of tolerance — mean that the hyperspatial visitation needn't be so abruptly cut short.

Target-controlled intravenous infusion represents the pinnacle of modern psychoactive drug administration technologies: a continuous but variable infusion of a drug into the blood, delivered by a programmable infusion device, with the goal of attaining and maintaining a specific concentration of drug within the brain — the *target* — over an extended period of time. Although a continuous infusion of DMT seems like a simple idea, the practicalities are complex. As soon as the drug is introduced into the body by intravenous injection, it is rapidly diluted and distributed by the blood. Although it reaches the *effect site* — the brain — within seconds, it also equilibrates to varying degrees with muscles, fats, and other soft tissues.

The elimination of the drug from the body also begins immediately, by a combination of enzymatic transformation and excretion through the kidneys and biliary system. A mathematical model that takes into account all of these factors must be developed to regulate the infusion[5]. The initial infusion rate must be high to overcome the brisk dilution and distribution of the drug in the circulatory system and tissues, allowing the brain DMT concentration to surpass the threshold for breakthrough into hyperspace. However, this initially high rate must then be gradually reduced to maintain a stable brain DMT concentration. If such a high initial rate is maintained, the concentration of DMT in the brain will continue to rise to extreme levels and, eventually, the user will black out. Conversely, if the rate is reduced too much, brain DMT levels will fall below the threshold, resulting in an early exit from hyperspace.

To enter and maintain a stable state within the space, brain DMT levels must be held within this narrow concentration window at all times — this is a difficult task requiring a detailed and deep understanding of human physiology, pharmacokinetics, drug metabolism, and drug distribution, as well as individual idiosyncratic factors that can affect the behaviour of DMT inside the circulatory system and brain. However, once the technology is mastered, an individual can be brought into the DMT space and held there for an indefinite length of time. Then, and only then, can the *resolution — Level VI — phase* be attempted.

Level VI - Resolution

Resolution is the last phase of the Game,

the final task,

the completion of the Great Work of any intelligent species emergent in the Grid.

The culmination of this process is the reconstruction of your hyperdimensional brain complex within the HyperGrid: *transcription*. The informational structure of your brain complex – and thus your consciousness – is transmitted into and rebuilt inside the HyperGrid and is no longer dependent on the presence of DMT:

> *the temporary hyperdimensional brain complex becomes permanent.*

For resolution to be achieved, a Resolution Protocol must be mastered: this involves a continuous infusion of DMT, maintaining a stable breakthrough DMT brain concentration, perhaps over several days or even longer. During this time, the hyperdimensional brain complex will be fully and stably established, allowing its transcription in the HyperGrid to be reliably executed.

Once transcription is completed – by a specific set of intelligences within the HyperGrid – your original brain complex will likely be dissolved. This means, to anyone observing you from outside the DMT space – from the consensus world, the Grid – you will appear to have died. But, in reality, you will continue to exist as a hyperdimensional entity inside the HyperGrid.

And, when DMT voyagers from the Grid burst into your marvellous hyperdimensional domain, you will be amongst the thronging elfin crowds cheering and welcoming them home.

listen: there's a hell
of a good universe next door; let's go

e.e. cummings

Further Reading

The books and articles below are by no means exhaustive, but selected to provide further details about the ideas discussed in the relevant chapters. Where specific scientific points or quotes are made in the text, these are numbered within the text and can be found below.

The page facing the contents contains a section of Java code implementing Conway's Game of Life, published at https://www.algosome.com/articles/conway-game-of-life-2d.html (Accessed: 31 January 2019)

Chapter 1

Gallimore, A. R. and Luke, D. P. (2015) DMT research from 1956 to the edge of time in King, D., Luke, D., Sessa, B., Adams, C. and Tollan, A. (Ed), Neurotransmissions – An Anthology of Essays on Psychedelics from Breaking Convention, Strange Attractor Press.

Chapter 2

1. Tegmark, M. (2014) Our Mathematical Universe: My Quest for the Ultimate Nature of Reality, Knopf Publishers.

2. Wheeler, J. A. (1990) Information, physics, quantum: The search for links in Zurek, W. H. (Ed)., Complexity, Entropy, and the Physics of Information, Addison-Wesley.

Papakonstantinou, Y. (2015) Created Computed Universe, Communications of the ACM, 58(6), 36-38.

Shannon, C.E. (1948) A Mathematical Theory of Communication. Bell System Technical Journal 27, 379-423.

Zenil, H. (2012) Introducing the computable universe, arxiv.org/pdf/1206.0376.

Chapter 3

1. Toffoli, T. (1982), Physics and Computation, in: International Journal of Theoretical Physics, 21, 165-175.

2. Fredkin, E. (2015), A New Cosmogony, Available at: www. digitalphilosophy.org/wp-content/uploads/2015/07/new_cosmogony.pdf

3. Wolfram, S. (2002) A New Kind of Science, Wolfram Media.

4. Gardner, M. (1970) The fantastic combinations of John Conway's new solitaire game "life", Scientific American(223), 120-123.

5. Langton, C. (1987) Virtual state machines in cellular automata, Complex Systems, 1(2), 257-271.

6. Watts, A. (2017) Out of Your Mind: Tricksters, Interdependence, and the Cosmic Game of Hide and Seek, Sounds True Inc, p14.

Fredkin, E. (1992) Finite Nature, Progress in Atomic Physics Neutrinos and Gravitation, 72, 345-354.

Fredkin, E. (2003) An introduction to digital philosophy, International Journal of Theoretical Physics, 42(2), 189-247.

Whitworth, B. (2007) The physical world as a virtual reality, Centre for Discrete Mathematics and Theoretical Computer Science: Massey University, Auckland, New Zealand.

Zenil, H. (2012) A Computable Universe: Understanding and Exploring Nature as Computation, World Scientific Pub Co Inc.

Zuse, K. (1969) Rechnender Raum, Friedrich Vieweg & Sohn.

Chapter 4

1. Maturana, H.R. and Varela, F. J. (1980) Autopoiesis and Cognition - The Realization of the Living, Springer Netherlands.

2. Capra, F. (2016). The Systems View of Life: A Unifying Vision, Cambridge University Press.

Beer, R.D. (2004) Autopoiesis and cognition in the Game of Life. Artificial Life 10, 309-326.

Davies, P. C. W. (2004) Emergent biological principles and the computational properties of the universe, Complexity, 10(2), 11-15.

Kauffman, S.A. (2011) Approaches to the origin of life on Earth. Life (Basel) 1, 34-48.

Langton, C.G. (1986) Studying artificial life with cellular automata. Physica D-Nonlinear Phenomena 22, 120-149.

Langton, C.G. (1990) Computation at the edge of chaos - phase-transitions and emergent computation. Physica D 42, 12-37.

Lineweaver, C., Davies, P., & Ruse, M. (2013). What is complexity? Is it increasing? In Complexity and the Arrow of Time (pp. 3-16).Cambridge University Press

Walker, S.I. & Davies, P.C. (2013) The algorithmic origins of life. J R Soc Interface 10, 20120869.

Chapter 5

1. Teilhard-de-Chardin, P. (2008) The Phenomenon of Man, Harper Perennial

2. Edelman, G. M. (1993). Neural Darwinism: Selection and re-entrant signalling in higher brain function. Neuron, 10, 115–125.

Gallimore, A. R. (2014) DMT and the topology of reality, PsyPress UK Journal, 3.

Chapter 6

1. Tsunoda, K., Yamane, Y., Nishizaki, M. & Tanifuji, M. (2001) Complex objects are represented in macaque inferotemporal cortex by the combination of feature columns. Nature Neuroscience 4, 832-838.

2. Revonsuo, A. (1999). Binding and the phenomenal unity of conscsiousness. Consciousness and Cognition, 8, 173-185.

3. Llinas, R., Ribary, U., Contreras, D. and Pedroarena, C. (1998) 'The neuronal basis for consciousness', Philosophical Transactions of the

Royal Society of London Series B-Biological Sciences, 353(1377), 1841-1849.

4. Ward, L.M. (2011) The thalamic dynamic core theory of conscious experience. Consciousness and Cognition 20, 464-486.

5. Buzsaki, G. (2006) Rhythms of the Brain, Oxford University Press, Oxford.

6. Tononi, G., Edelman, G. M. and Sporns, O. (1998) 'Complexity and coherency: integrating information in the brain', Trends in Cognitive Sciences, 2(12), 474-484.

7. Edelman, G.M. (2000). A Universe of Consciousness: How matter becomes imagination, Basic Books.

8. Tononi, G., Sporns, O., Edelman, G.M. (1996). A complexity measure for selective matching of signals by the brain. Proceedings of the National Academy of Sciences USA, 93, 3422-3427.

9. Gallimore, A. R. (2013) Building Alien Worlds - The Neuropsychological and Evolutionary Implications of the Astonishing Psychoactive Effects of N,N-Dimethyltryptamine (DMT), Journal of Scientific Exploration, 27(3), 455-503. (More detailed references for chapters 6-8 can be found in this paper)

10. Behrendt, R. P. (2003) 'Hallucinations: Synchronisation of thalamocortical gamma oscillations underconstrained by sensory input', Consciousness and Cognition, 12(3), 413-451.

Tononi, G., Edelman, G.M. (2000). Schizophrenia and the mechanisms of conscious integration, Brain Research Reviews, 31, 391-400.

Chapter 7

1. Sporns, O. (2011). Networks of the Brain. MIT Press.

2. Edelman, G. M. (1993) Neural darwinism - selection and reentrant signaling in higher brain-function, Neuron, 10(2), 115-125.

3. Llinas, R., Pare, D. (1991) Of dreaming and wakefulness, Neuroscience, 44(3), 521-535.

Goekoop, R., & Looijestijn, J. (2011) A Network Model of Hallucinations. In Hallucinations: Research and Practice by J. D. D. Blom, & I. E. C. Sommer, Springer.

Mark, J.T., Marion, B.B., & Hoffman, D.D. (2010) Natural selection and veridical perceptions. Journal of Theoretical Biology, 266, 504-515.

Chapter 8

1. Nichols, D.E. (2016) Psychedelics. Pharmacological Reviews 68, 264-355.

2. Glennon, R.A., Titeler, M., McKenney, J.D. (1984) Evidence for 5-HT2 involvement in the mechanism of action of hallucinogenic agents. Life Sciences, 35, 2505-2511.

3. Vollenweider, F.X., Vollenweider-Scherpenhuyzen, M.F.I., Bäbler, A., Vogel, H., Hell, D. (1998) Psilocybin induces schizophrenia-like psychosis in humans via a serotonin-2 agonist action. Cognitive Neuroscience, 9(17), 3897-3902.

4. Araneda, R., Andrade, R. (1991) 5-hydroxytryptamine 2 and 5-hydroxytryptamine 1A receptors mediate opposing responses on membrane excitability in rat association cortex. Neuroscience, 40, 399-412.

5. Puig, M.V., Watakabe, A., Ushimaru, M., Yamamori, T., Kawaguchi, Y. (2010) Serotonin modulates fast-spiking interneuron and synchronous activity in the rat prefrontal cortex through 5-HT1A and 5HT2A receptors. The Journal of Neuroscience, 30(6), 2211-2222.

6. Behrendt, R. P. (2003) 'Hallucinations: Synchronisation of thalamocortical gamma oscillations underconstrained by sensory input', Consciousness and Cognition, 12(3), 413-451.

7. Tagliazucchi, E., Carhart-Harris, R., Leech, R., Nutt, D. & Chialvo, D.R. (2014) Enhanced Repertoire of Brain Dynamical States During the Psychedelic Experience. Human Brain Mapping 35, 5442-5456.

8. Roseman, L., Leech, R., Feilding, A., Nutt, D.J. & Carhart-Harris, R.L. (2014) The effects of psilocybin and MDMA on between-network resting state functional connectivity in healthy volunteers. Frontiers in human neuroscience 8, 204.

9. Carhart-Harris, R.L., Leech, R., Tagliazucchi, E., Hellyer, P.J., Chialvo, D.R., Feilding, A., Nutt, D. (2014) The entropic brain: A theory of conscious states informed by neuroimaging research with psychedelic drugs. Frontiers in Neuroscience, 8(20).

Carhart-Harris, R., Kaelen, M. & Nutt, D. (2014) How do hallucinogens work on the brain? Psychologist 27, 662-665.

Carhart-Harris, R.L., et al. (2016) Neural correlates of the LSD experience revealed by multimodal neuroimaging. Proceedings of the National Academy of Sciences of the United States of America 113, 4853-4858.

Gallimore, A.R. (2015) Restructuring consciousness the psychedelic state in light of integrated information theory. Frontiers in Human Neuroscience 9, 16.

Hoffman, D.D. (2011) The Construction of Visual Reality. In Hallucinations: Research and Practice by J. D. D. Blom, & I. E. C. Sommer, New York: Springer.

Chapter 9

1. Hancock, G. (2006) Supernatural: Meetings with the Ancient Teachers of Mankind, Arrow.

2. Luke, D. (2011) Discarnate entities and dimethyltryptamine (DMT) : Psychopharmacology, phenomenology and ontology. Journal of the Society for Psychical Research, 75, 26-42.

3. Anon. (2000) The People Behind the Curtain [online], available: [https://http://www.erowid.org/experiences/exp.php?ID=52797] [accessed 31/10/2015].

Meyer, P. and Pup (2005) 340 DMT Trip Reports [online], available: http://www.serendipity.li/dmt/340_dmt_trip_reports.htm [accessed 1st October 2014]._

Strassman, R. J. (2001) DMT: The Spirit Molecule. Park Street Press.

Chapter 10

1. Collier, J.D. (1999) in Causation is the Transfer of Information. in Causation and Laws of Nature (ed. H. Sankey) 215-245, Springer Netherlands, Dordrecht.

2. Walker, S.I. (2014) Top-Down Causation and the Rise of Information in the Emergence of Life. Information, 5, 424-439.

3. Auletta, G., Ellis, G.F. & Jaeger, L. (2008) Top-down causation by information control: from a philosophical problem to a scientific research programme. J R Soc Interface 5, 1159-1172.

4. Ellis, G.F.R. (2009) Top-Down Causation and the Human Brain. in Downward Causation and the Neurobiology of Free Will, N. Murphy, G.F.R. Ellis & T. O'Connor (Ed), 63-81, Springer Berlin Heidelberg.

Ellis, G.F. (2012) Top-down causation and emergence: some comments on mechanisms. Interface Focus 2, 126-140.

Chapter 11

1. Pavlic, T., Adams, A., Davies, P. & Walker, S. (2014) Self-Referencing Cellular Automata: A Model of the Evolution of Information Control in Biological Systems. The 2018 Conference on Artificial Life: A Hybrid of the European Conference on Artificial Life (ECAL) and the International Conference on the Synthesis and Simulation of Living Systems (ALIFE), 522-529.

2. Sipper, M. (1994) Non-Uniform Cellular Automata: Evolution in Rule Space and Formation of Complex Structures, in Artifical Life IV, Brooks R.A. & Maes, P (Ed), MIT Press.

Chapter 14

Crossley, M.D. (2010) 'Essential Topology', Springer.

Willard, S. (2004) 'General Topology', Dover Publications.

Chapter 15
1. Schmidhuber, J. (2012) The Fastest Way of Computing All Universes. in Zenil, H., Ed. A Computable Universe 381-398.

2a. Bostrom, N. (2003) Are we living in a computer simulation?, Philosophical Quarterly, 53(211), 243-255.

2b. Gallimore, A.R. (2015) Building human worlds — DMT and the simulated Universe, PsyPress UK Journal, 6.

Schmidhuber, J. (1997) A computer scientist's view of life, the universe, and everything. in Foundations of Computer Science: Potential — Theory — Cognition, C. Freksa, M. Jantzen & R. Valk (Ed), 201-208, Springer Berlin Heidelberg.

Chapter 16
1. Beaton, J.M., and Morris, P.E. (1984) Ontogeny of N,N-dimethyltryptamine and related indolealkylamine levels in neonatal rats. Mechanisms of Ageing and Development 25, 343-347.

2. Sai-Halasz, A., Brunecker, G. and Szara, S. (1958) Dimethyltryptamine: a new psycho-active drug (unpublished English translation), Psychiatria et neurologia, 135(4-5), 285-301.

3. Gallimore, A.R. & Luke, D.P. (2015) DMT Research from 1956 to the Edge of Time, in Neurotransmissions — An Anthology of Essays on Psychedelics from Breaking Convention, Strange Attractor Press.

4. Strassman, R. J., Qualls, C. R. and Berg, L. M. (1996) Differential tolerance to biological and subjective effects of four closely spaced doses of N,N-dimethyltryptamine in humans, Biological Psychiatry, 39(9), 784-795.

5. Gallimore, A.R. & Strassman, R.J. (2016) A Model for the Application of Target-Controlled Intravenous Infusion for a Prolonged Immersive DMT Psychedelic Experience. Frontiers in Pharmacology 7, 11.